SpringerBriefs in Mol

Electrical and Magnetic Properties of Atoms, Molecules, and Clusters

More information about this series at http://www.springer.com/series/11647

Masayoshi Nakano

Excitation Energies and Properties of Open-Shell Singlet Molecules

Applications to a New Class of Molecules for Nonlinear Optics and Singlet Fission

Masayoshi Nakano
Department of Materials
Engineering Science
Osaka University
Toyonaka, Osaka
Japan

ISSN 2191-5407 ISSN 2191-5415 (electronic)
ISBN 978-3-319-08119-9 ISBN 978-3-319-08120-5 (eBook)
DOI 10.1007/978-3-319-08120-5

Library of Congress Control Number: 2014944342

Springer Cham Heidelberg New York Dordrecht London

Printed on acid-free paper

Springer is part of Springer Science+Business Media (www.springer.com)

To
My Wife Satomi, Daughter Makiko,
and Son Yoshiyuki

Preface

This book presents a novel viewpoint, "open-shell character," which is a quantum-chemically defined chemical index for bond nature, for excitation energies, and properties of open-shell singlet molecules, as well as its application to molecular design for two functionalities, i.e., nonlinear optics (NLO) and singlet fission.

Excitation energies and transition properties, e.g., transition moments and dipole moment differences, between two electronic states of a molecule are known to be understood by analyzing the primary configurations describing the two electronic states in the configuration interaction (CI) scheme. In the case of single excitation CI scheme, the ground state is unchanged, e.g., the Hartree–Fock ground state, while the excited states are usually described by a few singly excited configurations, e.g., HOMO (the highest occupied molecular orbital) → LUMO (the lowest unoccupied molecular orbital), which implies the ground-state Slater determinant with a substitution of HOMO by LUMO. In such a case, the excitation energies and transition properties are understood by the orbital energy gap and one-electron transition moment between the corresponding occupied and unoccupied MOs, respectively. Thus, various optical spectra and response phenomena can be described by using the single excitation picture. Although such a simple one-electron picture is preserved for weak electron-correlated systems, it is broken down for intermediate/strong electron-correlated systems, which are characterized by the nonzero open-shell character as explained later. The target systems in this book—open-shell singlet molecules—really belong to the class of intermediate/strong electron-correlated systems. The simplest model for such systems is a two-site diradical model with two electrons in two active orbitals [bonding (g) and antibonding (u) orbitals], where the open-shell (diradical) nature is, for instance, controlled by varying the interatomic distance. In the equilibrium bond length region, the system is regarded as a closed-shell system, while the diradical nature gradually increases by increasing the bond length, and finally the bond of two-site system is completely broken. Such a bond breaking is described by the doubly excited configuration from g to u, which becomes more significant with increasing the bond distance due to the decrease in the g–u gap. Namely, the weight of double excitation configurations in the ground-state wave function is a measure of "*electron correlation*," i.e., the degree of localization of each site, and thus represents the "*diradical character*." As expected from the correlation between these configurations in the two-site model, the excitation energies and properties are described as functions of diradical character in the ground state. This implies that the excitation energies and properties have the possibility of being controlled through tuning the diradical character in the ground state. In addition, the diradical character is closely related to the molecular architecture, MO picture (one-electron picture), aromaticity and so on since it is a chemical index of a bond nature. Namely, tuning diradical character in molecules is expected to be relatively easy because of the close relationships between the diradical character and the conventional chemical indices, which many chemists are familiar with. This fact describes a fundamental concept in this book, i.e., "diradical character view of various physico-chemical phenomena," which is thus useful for designing functional molecular systems. In this book, we present several molecular design principles for efficient NLO molecules as well as for singlet fission molecules based on the open-shell singlet molecules, where the intermediate open-shell character is revealed to be a key factor. The performances of these functionalities in intermediate open-shell singlet systems are demonstrated to surpass those in conventional closed-shell systems. In conclusion, the open-shell character viewpoint provides a new basis for comprehensive understanding of excitation energies and properties for a wide range of molecular systems including intermediate/strong electron-correlated systems as well as for constructing novel molecular design principles for highly efficient functional molecular systems.

First of all, I would like to thank Professor Dr. Kizashi Yamaguchi (Osaka University) for his introducing me to the wonderful field of open-shell systems during my doctoral studies in Graduate School of Engineering Science, Osaka University. I am also thankful to Prof. Dr. George Maroulis (University of Patras) for inviting me to give several lectures on the first part of this book in International Conference of Computational Methods in Sciences and Engineering (ICCMSE 2004–2011) held in Greece. I would like to thank Assistant Professor Dr. Ryohei Kishi, Associate Prof. Dr. Yasuteru Shigeta (Professor in University of Tsukuba, at present), Associate Prof. Dr. Hideaki Takahashi (Associate Professor in Tohoku University, at present), and JSPS postdoctoral fellow Dr. Shabbir Muhammad (Assistant Professor in King Khalid University, at present) in our laboratory for many valuable discussions and suggestions, and for supplying numerical calculation data. Furthermore, it is a pleasure to thank Dr. Takuya Minami (Showa Denko K.K.), Dr. Hitoshi Fukui (Sumitomo Chemical Company, Limited), and Dr. Kyohei Yoneda (Assistant Professor in Nara National College of Technology, at present), who worked as graduate students in our laboratory, for making numerical analysis, preparing many of the figures, and for valuable discussions. In particular, Dr. Takuya Minami has been an invaluable help with the work on singlet fission discussed in the latter part of this book. I would also like to acknowledge Prof. Dr. Benoît Champagne (University of Namur), Prof. Dr. Takashi Kubo (Osaka University), Dr. Kenji Kamada (AIST), Prof. Dr. Koji Ohta (Kyoto University), and Associate Prof. Dr. Frédéric Castet (Université Bordeaux 1), who are collaborators sharing the common interests in the open-shell singlet NLO and singlet fission molecules, for many comments, suggestions, and discussions on the subjects included in this book. In addition, we would like to thank Prof. Dr. Yoshito Tobe (Osaka University), Assistant Prof. Dr. Akihiro Shimizu (Kyoto University), Prof. Dr. Hiroshi Miyasaka (Osaka University), Prof. Dr. Yosuke Yamamoto (Hiroshima University), Prof. Dr. Manabu Abe (Hiroshima University), Prof. Dr. Akira Sekiguchi (University of Tsukuba), Prof. Dr. Hiroshi Ikeda (Osaka Prefecture University), and Prof. Dr. Takeaki Iwamoto (Tohoku University) for sharing with me their excellent idea and wonderful experimental results on various unique molecules. I also wish to thank Mr. Kotato Fukuda (a Doctor course student at Osaka University) for his careful checking of the manuscript, formulas, and so on. Finally, numerous students and colleagues have helped me by preparing many numerical data and analysis and by making useful suggestions. Although I will not recall everyone, this book could not have seen the light of day without his/her contribution.

Toyonaka, Osaka, March 2014 Masayoshi Nakano

Contents

Chapter 1
Introduction

Abstract The concept of "diradical character based design" for efficient functional substances is introduced using the dissociation process of a homodinuclear system. The diradical character, which is one of the quantum-chemically well-defined chemical indices and indicates the singlet open-shell nature, is employed for classification of arbitrary electronic structures into three categories, i.e., weak, intermediate and strong electron correlation regions. In this book, we present a simple relationship between diradical character and the ground/excited electronic structures, and illuminate that the systems in the intermediate diradical character region have the advantage of exhibiting highly efficient optoelectronic functionality. As examples, we show the diradical character based molecular design principles for highly efficient nonlinear optical (NLO) and singlet fission (SF) properties.

Keywords Diradical character · Open-shell singlet · Electron correlation · Excitation · Nonlinear optics · Singlet fission

1.1 Classification of Electronic Structure Based on the Diradical Character

There have been several reviews on the open-shell molecular systems, in particular, biradicals (or diradicals) [1–6]. Diradicals are defined as molecules involving two unpaired (odd) electrons, and provide a fundamental concept for chemical bond nature. Namely, the decrease in bond strength corresponds to the increase in diradical character. Thus, the diradical character is a key factor of determining chemical structures, reactivities and properties of molecules (Fig. 1.1a). The theoretical description of electronic structures of diradicals can be performed using the two-site model with two electrons in two frontier orbitals [1–8]. For example, the variation in diradical character can be well described by the dissociation of a dinuclear system as shown in Fig. 1.1b. In the spin-restricted (R) molecular orbital (MO) picture, the variation in diradical character is related to the variation in the

M. Nakano, *Excitation Energies and Properties of Open-Shell Singlet Molecules*,
SpringerBriefs in Electrical and Magnetic Properties of Atoms, Molecules, and Clusters,
DOI 10.1007/978-3-319-08120-5_1

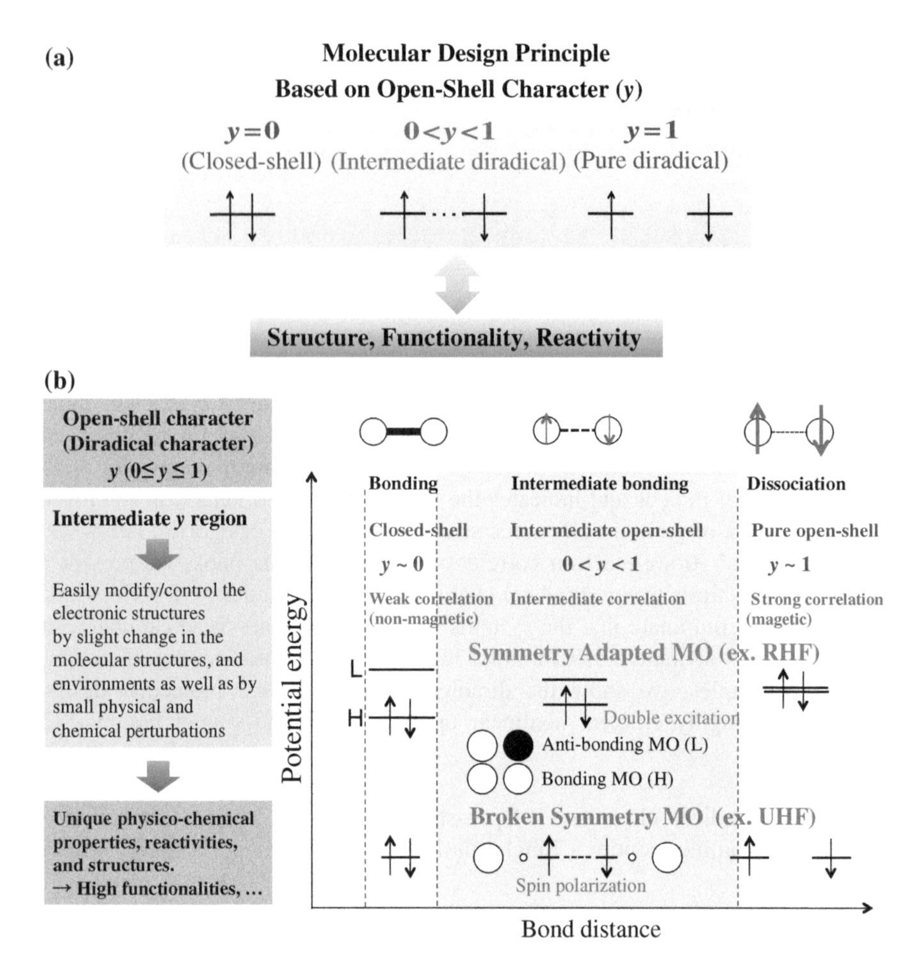

Fig. 1.1 Concept of diradical character based design principle (**a**). Variation in diradical character (y) is shown in the dissociation process of a dinuclear system (**b**). Potential energy curve and relationships between open-shell nature (from closed-shell to pure open-shell), bond nature and electron correlation are also shown together with the spin-restricted (symmetry-adapted) and unrestricted (broken-symmetry) molecular orbital (MO) descriptions of the electronic structures

energy gap between the highest occupied MO (HOMO) and the lowest unoccupied MO (LUMO): the diradical character tends to increase with the decrease in the HOMO-LUMO gap. Since the HOMO and LUMO possess the bonding and anti-bonding natures, respectively, the decrease in the HOMO–LUMO gap tends to cause the variation in bond nature from the stable bond regime to the bond dissociation regime through increasing the weight of doubly excitation configuration from HOMO to LUMO. In the stable bond limit, which is referred to as closed-shell, the diradical character disappears, while in the bond dissociation limit, the perfect diradical, which is referred to as pure open-shell, is obtained. The intermediate, or imperfect, diradical is referred to as “diradicaloid” (diradical-like)

[1–6]. To describe this continuous variation in the open-shell (diradical) nature of the two-electron two-orbital model, one can consider the use of the occupation number of the lowest unoccupied natural orbital (LUNO), which is equal to twice the weight of doubly excitation configuration in this case [9–13]. In the ground-state closed-shell molecules, all the occupation numbers are 2 (doubly occupied) or 0 (vacant), while in the open-shell singlet molecules, the NO occupation number can take an intermediate value between 0 and 2 depending on the diradical character. The occupation number of the LUNO is 0 in the case of a closed-shell molecule, while it increases with the increase in the diradical nature, and then approaches 1 when the system becomes a pure diradical. This implies that the diradical character, which is denoted by y hereafter ($0 \leq y \leq 1$), can be defined by the occupation number of the LUNO [9–13]. The physical and chemical meaning of diradical character is the degree of localization of two electrons on each site, that is, the degree of electron correlations. In other words, $q = 1 - y$ implies the "effective bond order": $q = 1$ (close-shell) and 0 (pure diradical) [10]. The exact solution of the two-electron two-orbital model with a minimal basis provides analytical expressions of the energies and wavefunctions of three singlet (S_0, S_1, S_2) and a triplet (T) states [7, 8]. In this book, we present the analytical expressions of excitation energies and properties (transition moments and dipole moment differences) using the diradical character (y) [8]. These expressions are useful for revealing the chemical concepts for excitation energies and properties based on the ground-state bond nature (diradical character). The excitation energies and properties as the function of the diradical character are explained in the symmetric (Chap. 2) and asymmetric (Chap. 3) two-site diradical model systems.

As applications of the diradical character concept, we propose two functional molecular design principles: one is for highly efficient nonlinear optical (NLO) open-shell molecules [8, 14–20] (Chap. 4) and the other is for feasibility conditions of efficient singlet fission molecules [21] (Chap. 5). These two design principles do not only shed light on the new aspect of "diradical character" as a controlling factor for functionalities, but also provide a novel class of functional systems, that is, "open-shell singlet molecules", whose functionalities could by far outstrip those of conventional closed-shell molecules.

1.2 Diradical Character Based Design Principles for Nonlinear Optical Properties and Singlet Fission

The microscopic origin of nonlinear optical (NLO) phenomena is the nonlinear electronic polarization of molecules through the nonlinear molecule—laser field interaction, which is characterized by the nonlinear optical polarizabilities, that is, hyperpolarizabilities [22–24], which are intrinsic molecular properties and can be described by spectroscopic quantities (excitation energies, transition moments and dipole moment differences). Several NLO phenomena including frequency coversion, field induced change in refractive index and multi-photon absorption/emission, are expected to be utilized in future optoelectronic and photonic devices for laser frequency modulation and ultrafast all optical switching, as well as in three-dimensional (3D) microfabrication, 3D optical data storage, and NLO spectroscopy [25–31]. The enhancement of hyperpolarizabilities and the development of their control schemes are indispensable for realizing these NLO-based applications. So far, a wide variety of organic and inorganic molecular systems have been explored both theoretically and experimentally [22–24]. Although conventional NLO substances are generally inorganic crystals such as lithium niobate ($LiNbO_3$) and potassium dihydrogen phosphate (KH_2PO_4), which exhibit second harmonic generation (SHG) effect, since the 1990s, π-electron conjugated organic molecular systems have emerged as promising candidates for highly efficient NLO applications [22–24] due to the larger optical nonlinearities and faster optical responses as well as to the lower driving voltages, diversity of molecular design, and potentially lower processing cost. On the basis of theoretical and experimental studies, various molecular design guidelines have been proposed for giving large hyperpolarizabilities and for controlling their features (sign and/or magnitudes) [25–31]. However, most of the investigations of NLO properties have been limited to closed-shell neutral systems though pioneering studies have been carried out on charged and/or radical compounds [32–34]. In this book, we present a novel design principle for the NLO properties of open-shell singlet systems with even number of electrons, e.g., singlet diradicals/multi-radicals, as well as some examples of real molecules designed by this principle.

Another phenomenon—Singlet fission (SF) is the photochemical process, where a singlet exciton splits into two triplet excitons [21]. This discovered in the 1960s [35–37], but this has recently attracted a great deal of interests both from science and engineering communities due to its potential application in organic solar cells, in which SF has a possibility of overcoming the conventional maximum limit of photoelectric conversion efficiency [38–41]. This fascinating property originates form the multiple carrier generation from split triplet excitons [38] as well as from the longer lifetime of triplet exciton than that of singlet one, which elongates the exciton diffusion length [42]. At the present time, although the number of molecules which exhibit SF continues to grow rapidly, more sophisticated molecular design guidelines for energetically efficient SF is necessary for realizing a high yield multielectron generating system for real photovoltaic application. In general, most of molecules have relatively large first triplet excitation energies [$E(\mathrm{T}_1)$], which result in an endoergic SF process, the decrease in $E(\mathrm{T}_1)$ is an important issue for the SF molecular design. Based on the careful observation and analysis, Michl et al. have proposed the energy level matching conditions of SF, which should be satisfied by an isolated single molecule [21],

$$2E(\mathrm{T}_1) - E(\mathrm{S}_1) \sim 0 \quad \text{or} \quad < 0, \quad \text{condition (i)} \tag{1.1}$$

$$2E(\mathrm{T}_1) - E(\mathrm{T}_2) < 0. \quad \text{condition (ii)} \tag{1.2}$$

Equation (1.1) is required to split singlet exciton (S_1) into two triplet excitons (T_1), and (1.2) is done to suppress the triplet-triplet annihilation, which generates a singlet ground state (S_0) and a higher triplet state (T_2) from split two triplet excitons (T_1). Although these conditions ignore the effects of intermolecular interaction, nonadiabatic transition between potential energy surfaces due to the electron–vibration coupling, and geometry change, in real SF process [21], these conditions are known to be quite useful for prescreening of efficient SF molecules. Indeed, Michl et al. have proposed two kinds of candidate systems based on these conditions [21, 43–45]: (a) "alternant hydrocarbons" with large exchange integral K_{HL} concerning the HOMO and LUMO, which approximately describes the energy difference between S_1 and T_1 when both states are characterized by the HOMO → LUMO one-electron transition, and (b) "diradicaloids", which are expected to have small energy differences between S_0 and T_1. As mentioned earlier, the energies of these states are predicted to be correlated to the diradical character in the ground state. This suggests that a more unified design principle could be constructed from the diradical character viewpoint. In Chap. 5, we present a multiple diradical character based design principle deduced from the full configuration interaction calculations of tetraradical model systems [46], and demonstrate the performance of this principle by designing model and real molecular systems as well as quantum chemical calculations of their diradical characters and excitation energies.

In general, the molecular design principle of structure, functionality and reactivity should be able to be easily applied to a broad class of compounds having various architectures and size, as well as involving arbitrary atom species, and it should also have close relations to the conventional chemical concepts, e.g., π-conjugation, aromaticity/antiaromaticity, electronegativity, charge transfer, and so on, which enable the chemical experimentalists to understand the principle without difficulty and then to apply it to realistic molecular systems that they are interested in. On the other hand, such concepts should also be explicitly defined based on the simple but essential models concerning the target phenomena. In this regard, the "open-shell character" or "diradical character" is one of the chemical indices for effective chemical bonds, which is a measure of electron correlation in other words. The open-shell character is a quantity in the ground state of the systems, while it is closely related to the several excited states through the electron correlation. This implies that the easily obtained open-shell character could be employed to predict and/or to design the physico-chemical properties concerning the ground and excited states, which are difficult to be evaluated in general. This is the fundamental concept of this book as shown in Fig. 1.1. Thus, we focus on the open-shell singlet systems with a wide range of open-shell characters. There have already been known to be a variety of open-shell singlet molecules including polycyclic aromatic hydrocarbons (PAHs) like graphenes [47, 48], multiple metal-metal bonded systems [49–51], and heavier main-group compounds [5] and so on. Some of their open-shell singlet ground states are characterized by intermediate/strong correlation between anti-parallel spins, so that such systems belong to the intermediate/strong electron correlation regime. These correlated electronic structures generally tend to be sensitive to external chemical and physical perturbations, e.g., slight structural changes, donor/acceptor substitutions, intermolecular interactions, spin/charge state changes, solvent effects, and the application of electromagnetic fields. As mentioned earlier, such correlations also determine the features of the excited states because these states are closely correlated with each other, resulting in tuning the relative weights of ionic and covalent (diradical) contributions [1–8]. From these facts, the chemical structures, reactivities, and physico-chemical properties of the ground and excited states are expected to be strongly influenced by slight external perturbations as compared to the closed-shell systems, i.e., weak correlated systems [8]. Namely, the controllability and responsibility of such properties in the open-shell singlet systems, i.e, intermediate/strong electron-correlated systems, are expected to outstrip those of conventional closed-shell systems. In this book, we present the fundamentals of this correlation together with two applications to design of functional molecules, i.e., open-shell NLO and singlet fission molecules, and thus propose the "diradicalology" concept [52], which allows us to explore the origins and/or to create/control the electron-correlation-driven physico-chemical phenomena as a function of the diradical characters in the broad fields of physics, chemistry and biology.

References

1. L. Salem, C. Rowland, Angew. Chem., Int. Ed. **11**, 92 (1972)
2. W.T. Borden (ed.), *Diradicals* (Wiley, New York, 1982)
3. V. Bonaic-Koutecky, J. Koutecky, J. Michl, Angew. Chem. Int. Ed. **26**, 170 (1987)
4. A. Rajca, Chem. Rev. **94**, 871 (1994)
5. F. Breher, Coord. Chem. Rev. **251**, 1007 (2007)
6. M. Abe, Chem. Rev. **113**, 7011 (2013)
7. C.J. Calzado, J. Cabrero, J.P. Malrieu, R. Caballol, J. Chem. Phys. **116**, 2728 (2002)
8. M. Nakano, R. Kishi, S. Ohta et al., Phys. Rev. Lett. **99**, 033001 (2007)
9. E.F. Hayes, A.K.Q. Siu, J. Am. Chem. Soc. **93**, 2090 (1971)
10. K. Yamaguchi, in *Self-Consistent Field: Theory and Applications*, ed. by R. Carbo, M. Klobukowski (Elsevier: Amsterdam, 1990), p. 727
11. M. Head-Gordon, Chem. Phys. Lett. **372**, 508 (2003)
12. K. Kamada, K. Ohta et al., J. Phys. Chem. Lett. **1**, 937 (2010)
13. M. Nakano et al. Theor. Chem. Acc. **130**, 711 (2011); *erratum* **130**, 725
14. M. Nakano, R. Kishi et al., J. Phys. Chem. A **109**, 885 (2005)
15. M. Nakano, R. Kishi, S. Ohta et al., J. Chem. Phys. **125**, 074113 (2006)
16. M. Nakano, K. Yoneda et al, J. Chem. Phys. **131**, 114316 (2009)
17. M. Nakano, B. Champagne et al., J. Chem. Phys. **133**, 154302 (2010)
18. M. Nakano, T. Minami et al., J. Phys. Chem. Lett. **2**, 1094 (2011)
19. M. Nakano, T. Minami et al., J. Chem. Phys. **136**, 024315 (2012)
20. M. Nakano et al., J. Chem. Phys. **138**, 244306 (2013)
21. M.J. Smith, J. Michl, Chem. Rev. **110**, 6891 (2010)
22. D. Burland (ed.), Special Issue on Optical Nonlinearities in Chemistry. Chem. Rev. **94**, 1–278 (1994)
23. H.S. Nalwa and S. Miyata (eds.), *Nonlinear Optics of Organic Molecules and Polymers* (CRC, Boca Raton, FL, 1997)
24. H.S. Nalwa (ed.), *Handbook of Advanced Electronic and Photonic Materials and Devices*, vol. 9, (Academic Press, New York, 2001)
25. D.A. Pathenopoulos, P.M. Rentzepis, Science **245**, 893 (1989)
26. B.H. Cumpston, S.P. Ananthavel et al., Nature **398**, 51 (1999)

27. W.R. Dichtel, J.M. Serin et al., J. Am. Chem. Soc. **126**, 5380 (2004)
28. S. Kawata, H.-B. Sun, T. Tanaka, K. Takada, Nature **412**, 697 (2001)
29. W. Zhou, S.M. Kuebler et al., Science **296**, 1106 (2002)
30. M. Albota, D. Beljonne, J.-L. Brédas, J.E. Ehrlich, Science **281**, 1653 (1998)
31. F. Terenziani, C. Katan et al., Adv. Mater. Weinheim: Ger. **20**, 4641 (2008)
32. M. Nakano, K. Yamaguchi, Chem. Phys. Lett. **206**, 285 (1993)
33. M. Nakano, I. Shigemoto, S. Yamada, K. Yamaguchi, J. Chem. Phys. **103**, 4175 (1995)
34. M. Nakano, H. Nagao, K. Yamaguchi, Phys. Rev. A **55**, 1503 (1997)
35. S. Singh, W. Jones, J.W. Siebrand, B.P. Stoicheff, W.G. Schneider, J. Chem. Phys. **42**, 330 (1965)
36. N. Geacintov, M. Pope et al., Phys. Rev. Lett. **22**, 593 (1969)
37. R.E. Merrifield, P. Avakian et al., Chem. Phys. Lett. **3**, 386 (1969)
38. M.C. Hanna, A.J. Nozik, J. Appl. Phys. **100**, 074510 (2006)
39. I. Paci, J.C. Johnson et al., J. Am. Chem. Soc. **128**, 16546 (2006)
40. P.J. Jadhav, A. Mohanty et al., Nano. Lett. **11**, 1495 (2011)
41. A. Rao, M.W. Wilson et al., J. Am. Chem. Soc. **132**, 12698 (2010)
42. H. Najafov, B. Lee et al., Nat. Mater. **9**, 938 (2010)
43. E.C. Greyson, J. Vura-Weis et al., J. Phys. Chem. B **114**, 14168 (2010)
44. P.M. Zimmerman et al., J. Am. Chem. Soc. **133**, 19944 (2011)
45. P.E. Teichen et al., J. Phys. Chem. B **116**, 11473 (2012)
46. T. Minami, M. Nakano, J. Phys. Chem. Lett. **3**, 145 (2012)
47. C. Lambert, Angew. Chem., Int. Ed. **50**, 1756 (2011)
48. Z. Sun, J. Wu, J. Mater. Chem. **22**, 4151 (2012)
49. M. Nishino et al., J. Phys. Chem. A **101**, 705 (1997)
50. H. Fukui et al., J. Phys. Chem. Lett. **2**, 2063 (2011)
51. H. Fukui et al., J. Phys. Chem. A **116**, 5501 (2012)
52. M. Nakano et al., Int. J. Quant. Chem. **113**, 585 (2013)

Chapter 2
Electronic Structures of Symmetric Diradical Systems

Abstract In general, the electronic structures of a molecular system is characterized by using the "diradical character", which is well defined in quantum chemistry and implies a chemical index of a bond nature. In this chapter, we present analytical expressions for electronic energies and wavefunctions of the ground- and excited states as well as for the excitation energies and transition properties based on symmetric two-site diradical models with different diradical characters using the valence configuration interaction method.

Keywords Symmetric diradical system · Diradical character · Excitation energy · Transition moment · Valence configuration interaction

2.1 Symmetric Diradical Model Using the Valence Configuration Interaction Method

In this section, we consider a symmetric two-site diradical molecular model, $\dot{\mathrm{A}} - \dot{\mathrm{B}}$ with two electrons in two active orbitals, which can present the essential features of the electronic structures of general diradical molecules including open-shell polycyclic aromatic hydrocarbons (PAHs) [1, 2], transition-metal dinuclear systems [3–7] and so on. The symmetry-adapted bonding (g) and anti-bonding (u) molecular orbitals (MOs), which are the natural orbitals (NOs) obtained from the spin-unrestricted, i.e., broken-symmetry (BS), solutions like the spin-unrestricted Hartree-Fock (UHF) solution, are described using the atomic orbitals (AO), χ_{A} and χ_{B} (which are mutually nonorthogonal and have an overlap S_{AB}):

$$\begin{aligned} g(x) &= \frac{1}{\sqrt{2(1+S_{\mathrm{AB}})}}[\chi_{\mathrm{A}}(x)+\chi_{\mathrm{B}}(x)], \quad \text{and} \\ u(x) &= \frac{1}{\sqrt{2(1-S_{\mathrm{AB}})}}[\chi_{\mathrm{A}}(x)-\chi_{\mathrm{B}}(x)]. \end{aligned} \tag{2.1.1}$$

M. Nakano, *Excitation Energies and Properties of Open-Shell Singlet Molecules*,
SpringerBriefs in Electrical and Magnetic Properties of Atoms, Molecules, and Clusters,
DOI 10.1007/978-3-319-08120-5_2

The localized natural orbital (LNO) is defined as [8, 9]

$$a(x) \equiv \frac{1}{\sqrt{2}}[g(x) + u(x)] \approx \chi_A(x), \quad \text{and}$$
$$b(x) \equiv \frac{1}{\sqrt{2}}[g(x) - u(x)] \approx \chi_B(x). \tag{2.1.2}$$

which are well localized on one site (A or B), while have generally small tails on the other site, satisfying the orthogonal condition, $\langle a|b\rangle = 0$. In the dissociation limit ($S_{AB} = 0$), the LNOs are apparently identical with the AOs. Using these two types of basis sets, we describe the singlet BS MO solution. The BS MOs for the α and β spins (referred to as ψ_H^α and ψ_H^β, respectively) are described using symmetry-adapted MOs, g and u, by [10–12]

$$\psi_H^\alpha = (\cos\theta)g + (\sin\theta)u, \quad \text{and} \quad \psi_H^\beta = (\cos\theta)g - (\sin\theta)u. \tag{2.1.3}$$

where θ is a mixing parameter of g and u and takes a value between 0 and $\pi/4$. In another way using the LNOs, a and b, the BS MOs are given by [10–12]

$$\psi_H^\alpha = (\cos\omega)a + (\sin\omega)b, \quad \text{and} \quad \psi_H^\beta = (\cos\omega)b + (\sin\omega)a, \tag{2.1.4}$$

where $\omega (0 \le \omega \le \pi/4)$ is a mixing parameter of these LNOs ($\approx$AOs). In the case of $\theta = 0 (\omega = \pi/4)$, the BS MOs are reduced to the symmetry-adapted MO, $\psi_H^\alpha = \psi_H^\beta = g$. In contrast, in the case of $\theta = \pi/4 (\omega = 0)$, the BS MOs correspond to the LNOs, $\psi_H^\alpha = a$ and $\psi_H^\beta = b$. Thus, we can consider the two limits: (i) weak correlation limit (MO limit) at $\theta = 0 (\omega = \pi/4)$, giving symmetry-adapted closed-shell MO (g), and (ii) strong correlation limit [valence bond (VB) limit] at $\theta = \pi/4 (\omega = 0)$, giving LNOs ($a$ and b). Namely, the different orbitals for different spins (DODS) MOs represented by Eqs. (2.1.3) and (2.1.4) can describe both weak and strong correlation limits, i.e., MO and VB limits, as well as the intermediate correlation regime [10–12].

Let us consider a symmetric two-site diradical system with two electrons in two orbitals, g and u (a and b) and the z-component of spin angular momentum $M_s = 0$ (singlet and triplet). For $M_s = 0$, there are two neutral and ionic determinants:

$$|a\bar{b}\rangle \equiv |\text{core } a\bar{b}\rangle, \quad |b\bar{a}\rangle \equiv |\text{core } b\bar{a}\rangle \quad \text{(neutral)}, \tag{2.1.5a}$$

and

$$|a\bar{a}\rangle \equiv |\text{core } a\bar{a}\rangle, \quad \text{and} \quad |b\bar{b}\rangle \equiv |\text{core } b\bar{b}\rangle \quad \text{(ionic)}, \tag{2.1.5b}$$

where core denotes the closed-shell inner orbitals, and the upper- and non-bar indicate β and α spins, respectively. The electronic Hamiltonian H (in atomic units (a.u.), $\hbar = m = \mathrm{e} = 1$) for this model system is represented by

$$H = -\frac{1}{2}\sum_{i=1}^{N}\nabla_i^2 - \sum_{i=1}^{N}\sum_{A=1}^{2}\frac{Z_A}{r_{iA}} + \sum_{i=1}^{N}\sum_{j>i}^{N}\frac{1}{r_{ij}} = \sum_{i=1}^{N}h(i) + \sum_{i=1}^{N}\sum_{j>i}^{N}\frac{1}{r_{ij}}. \tag{2.1.6}$$

The valence configuration interaction (VCI) matrix of this Hamiltonian using the LNO basis takes the form [8, 9]:

$$\begin{pmatrix} \langle a\bar{b}|H|a\bar{b}\rangle & \langle a\bar{b}|H|b\bar{a}\rangle & \langle a\bar{b}|H|a\bar{a}\rangle & \langle a\bar{b}|H|b\bar{b}\rangle \\ \langle b\bar{a}|H|a\bar{b}\rangle & \langle b\bar{a}|H|b\bar{a}\rangle & \langle b\bar{a}|H|a\bar{a}\rangle & \langle b\bar{a}|H|b\bar{b}\rangle \\ \langle a\bar{a}|H|a\bar{b}\rangle & \langle a\bar{a}|H|b\bar{a}\rangle & \langle a\bar{a}|H|a\bar{a}\rangle & \langle a\bar{a}|H|b\bar{b}\rangle \\ \langle b\bar{b}|H|a\bar{b}\rangle & \langle b\bar{b}|H|b\bar{a}\rangle & \langle b\bar{b}|H|a\bar{a}\rangle & \langle b\bar{b}|H|b\bar{b}\rangle \end{pmatrix} = \begin{pmatrix} 0 & K_{ab} & t_{ab} & t_{ab} \\ K_{ab} & 0 & t_{ab} & t_{ab} \\ t_{ab} & t_{ab} & U & K_{ab} \\ t_{ab} & t_{ab} & K_{ab} & U \end{pmatrix}, \tag{2.1.7}$$

where the energy of the neutral determinant, $\langle a\bar{b}|H|a\bar{b}\rangle = \langle b\bar{a}|H|b\bar{a}\rangle$, is taken as the energy origin (0). U denotes the difference between on- and neighbor-site Coulomb repulsions $[U \equiv U_{aa} - U_{bb} = (aa|aa) - (bb|bb)]$. K_{ab} is a direct exchange integral $[K_{ab} = (ab|ba) \geq 0]$, and t_{ab} is a transfer integral [$t_{ab} = \langle a\bar{b}|H|b\bar{b}\rangle = \langle a|f|b\rangle \leq 0$, where f is the Fock operator in the LNO representation] [9]. Each matrix element is derived as follows.

$$\begin{aligned} \langle a\bar{b}|H|a\bar{b}\rangle &= \langle a\bar{b}|\sum_{i=1}^{N}h(i)|a\bar{b}\rangle + \langle a\bar{b}|\sum_{i=1}^{N}\sum_{j>i}^{N}\frac{1}{r_{ij}}|a\bar{b}\rangle \\ &= \langle a|h|a\rangle + \langle b|h|b\rangle + \sum_{c}^{\text{core}}\langle c|h|c\rangle + \frac{1}{2}\langle a\bar{b}||a\bar{b}\rangle \\ &\quad + \frac{1}{2}\langle \bar{b}a||\bar{b}a\rangle + \frac{1}{2}\sum_{c}^{\text{core}}\langle ac||ac\rangle + \frac{1}{2}\sum_{c}^{\text{core}}\langle ca||ca\rangle \\ &\quad + \frac{1}{2}\sum_{c}^{\text{core}}\langle \bar{b}c||\bar{b}c\rangle + \frac{1}{2}\sum_{c}^{\text{core}}\langle c\bar{b}||c\bar{b}\rangle + \frac{1}{2}\sum_{c}^{\text{core}}\sum_{c'}^{\text{core}}\langle cc'||cc'\rangle \\ &= \langle a|h|a\rangle + \langle b|h|b\rangle + \frac{1}{2}\langle a\bar{b}||a\bar{b}\rangle + \frac{1}{2}\langle \bar{b}a||\bar{b}a\rangle + (\text{core}) \\ &= (a|h|a) + (b|h|b) + \frac{1}{2}\{(aa|bb) + (bb|aa)\} + (\text{core}) \\ &= 2h_{aa} + U_{ab} + (\text{core}), \end{aligned} \tag{2.1.8a}$$

where all the terms concerning the core are denoted by "(core)", and $h_{aa}[\equiv(a|h|a)] = h_{bb}$ due to the symmetry of the model system. Similarly, we obtain

$$\langle b\bar{a}|H|b\bar{a}\rangle = 2h_{aa} + U_{ab} + (\text{core}). \tag{2.1.8b}$$

Since these energies are defined as the energy origin $[2h_{aa} + U_{ab} + (\text{core}) = 0]$, the energies of ionic determinants are represented by

$$\begin{aligned}\langle a\bar{a}|H|a\bar{a}\rangle = \langle b\bar{b}|H|b\bar{b}\rangle &= 2h_{aa} + U_{aa} + (\text{core}) \\ &= U_{aa} - U_{ab} = U,\end{aligned} \tag{2.1.8c}$$

where we use the relations: $\sum_c^{\text{core}} \langle ac||ac\rangle = \sum_c^{\text{core}} \langle bc||bc\rangle$, $h_{aa} = h_{bb}$ and $U_{aa} = U_{bb}$, which come from the symmetry of the present system.

For off-diagonal elements, we obtain

$$\langle a\bar{b}|H|b\bar{a}\rangle = \langle b\bar{a}|H|a\bar{b}\rangle = \langle a\bar{b}||b\bar{a}\rangle = (ab|ba) = K_{ab}, \tag{2.1.8d}$$

$$\begin{aligned}\langle a\bar{b}|H|a\bar{a}\rangle = \langle a\bar{a}|H|a\bar{b}\rangle &= \langle \bar{b}|h|\bar{a}\rangle + \langle \bar{b}a||\bar{a}a\rangle \\ &+ \sum_c^{\text{core}} \langle \bar{b}c||\bar{a}c\rangle = \langle \bar{b}|\hat{f}|\bar{a}\rangle = t_{ab},\end{aligned} \tag{2.1.8e}$$

$$\begin{aligned}\langle a\bar{b}|H|b\bar{b}\rangle = \langle b\bar{b}|H|a\bar{b}\rangle &= \langle b\bar{a}|H|a\bar{a}\rangle = \langle a\bar{a}|H|b\bar{a}\rangle \\ &= \langle b\bar{a}|H|b\bar{b}\rangle = \langle b\bar{b}|H|b\bar{a}\rangle = t_{ab},\end{aligned} \tag{2.1.8f}$$

and

$$\langle a\bar{a}|H|b\bar{b}\rangle = \langle b\bar{b}|H|a\bar{a}\rangle = \langle a\bar{a}||b\bar{b}\rangle = (ab|ab) = K_{ab}, \tag{2.1.8g}$$

where $t_{ab} = t_{ba}$ is used. By diagonalizing the CI matrix of Eq. (2.1.7), the four solutions are obtained as follows [8].

(i) Neutral triplet state (u symmetry)

$$|T_{1u}\rangle = \frac{1}{\sqrt{2}}\left(|a\bar{b}\rangle - |b\bar{a}\rangle\right) \text{ with energy } {}^3E_{1u} = -K_{ab}. \tag{2.1.9}$$

This triple state consists of only neutral determinants and is pure diraidcal.

(ii) Ionic singlet state (u symmetry)

$$|S_{1u}\rangle = \frac{1}{\sqrt{2}}\left(|a\bar{a}\rangle - |b\bar{b}\rangle\right) \text{ with energy } {}^1E_{1u} = U - K_{ab}. \quad (2.1.10)$$

(iii) Lower singlet state (g symmetry)

$$|S_{1g}\rangle = \kappa\left(|a\bar{b}\rangle + |b\bar{a}\rangle\right) + \eta\left(|a\bar{a}\rangle + |b\bar{b}\rangle\right), \quad (2.1.11a)$$

where $2(\kappa^2 + \eta^2) = 1$ and $\kappa > \eta > 0$. The energy is

$${}^1E_{1g} = K_{ab} + \frac{U - \sqrt{U^2 + 16t_{ab}^2}}{2}. \quad (2.1.11b)$$

In this state, $2\kappa^2$ and $2\eta^2$ represent the weight of neutral and ionic contributions, respectively, and the weight of neutral determinant is larger than that of ionic one.

(iv) Higher singlet state (g symmetry)

$$|S_{2g}\rangle = -\eta\left(|a\bar{b}\rangle + |b\bar{a}\rangle\right) + \kappa\left(|a\bar{a}\rangle + |b\bar{b}\rangle\right), \quad (2.1.12a)$$

where $2(\kappa^2 + \eta^2) = 1$ and $\kappa > \eta > 0$. The energy is

$${}^1E_{2g} = K_{ab} + \frac{U + \sqrt{U^2 + 16t_{ab}^2}}{2}. \quad (2.1.12b)$$

In this state, $2\kappa^2$ and $2\eta^2$ represent the weight of ionic and neutral contributions, respectively, and the weight of ionic determinant is larger than that of neutral one.

Here, we obtain [13]

$$\kappa = \frac{1}{2}\sqrt{1 + \frac{U}{\sqrt{U^2 + 16t_{ab}^2}}} \quad \text{and}$$
$$\eta = \frac{2|t_{ab}|}{\sqrt{\left(U + \sqrt{U^2 + 16t_{ab}^2}\right)\sqrt{U^2 + 16t_{ab}^2}}}. \quad (2.1.13)$$

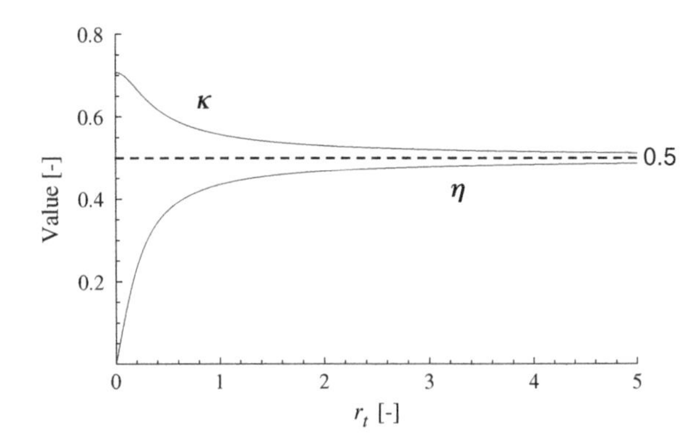

Fig. 2.1 Variations of κ and η with respect to r_t

If we define r_t as $r_t \equiv |t_{ab}/U|$, κ and η are rewritten as [13],

$$\kappa = \frac{1}{2}\sqrt{1 + \frac{1}{\sqrt{1 + 16r_t^2}}} \quad \text{and} \quad \eta = \frac{2r_t}{\sqrt{\left(1 + \sqrt{1 + 16r_t^2}\right)\sqrt{1 + 16r_t^2}}}. \tag{2.1.14}$$

A large t_{ab} and a small U, which lead to a large r_t, correspond to the ease of the electron transfer between sites A and B. Figure 2.1 shows the variations of κ and η with respect to r_t. As decreasing r_t, the coefficient (κ) of the neutral determinant increases toward $1/\sqrt{2}$ at $r_t = 0$, while that (η) of the ionic determinant decreases toward 0 at $r_t = 0$. This indicates that the mobility of electrons between sites A and B governs the relative neutral (covalent) and ionic natures of the state, resulting in determining the diradical nature.

2.2 Diradical Character of Symmetric Systems

2.2.1 Diradical Character in the VCI Model

The ground and excited states are also described using the symmetry-adapted MOs, g and u, [Eq. (2.1.1)] as

$$g(x) = \frac{1}{\sqrt{2}}[a(x) + b(x)] \quad \text{and} \quad u(x) = \frac{1}{\sqrt{2}}[a(x) - b(x)]. \tag{2.2.1}$$

We here employ the following determinants for $M_s = 0$ as an alternative basis to the LNO basis:

$\lvert g\bar{g}\rangle$	ground-state configuration with two electrons in HOMO
$\lvert g\bar{u}\rangle$ and $\lvert u\bar{g}\rangle$	HOMO to LUMO singly excited configuration
$\lvert u\bar{u}\rangle$	HOMO to LUMO doubly excited configuration

By employing these bases, we can obtain four wavefunctions for the z-component of spin angular momentum, $M_s = 0$, which are equivalent to those obtained by the VCI method using the LNOs [Eqs. (2.1.9), (2.1.10), (2.1.11a) and (2.1.12a)] [13]:

(i) Neutral triplet state (u symmetry)

$$|T_{1u}\rangle = \frac{1}{\sqrt{2}}(|g\bar{u}\rangle - |u\bar{g}\rangle) \tag{2.2.2}$$

(ii) Ionic singlet state (u symmetry)

$$|S_{1u}\rangle = \frac{1}{\sqrt{2}}(|g\bar{u}\rangle + |u\bar{g}\rangle) \tag{2.2.3}$$

(iii) Lower singlet state (g symmetry)

$$|S_{1g}\rangle = \xi|g\bar{g}\rangle - \zeta|u\bar{u}\rangle \tag{2.2.4}$$

(iv) Higher singlet state (g symmetry)

$$|S_{2g}\rangle = \zeta|g\bar{g}\rangle + \xi|u\bar{u}\rangle \tag{2.2.5}$$

By substituting Eq. (2.2.1) into Eq. (2.2.4), we obtain

$$\begin{aligned}|S_{1g}\rangle = \xi|g\bar{g}\rangle - \zeta|u\bar{u}\rangle &= \frac{1}{2}(\xi+\zeta)\left(|a\bar{b}\rangle + |b\bar{a}\rangle\right) \\ &+ \frac{1}{2}(\xi-\zeta)\left(|a\bar{a}\rangle + |b\bar{b}\rangle\right).\end{aligned} \tag{2.2.6}$$

By comparing the coefficients between Eqs. (2.1.11a) and (2.2.6), we obtain

$$\kappa = \frac{1}{2}(\xi+\zeta) \quad \text{and} \quad \eta = \frac{1}{2}(\xi-\zeta) \tag{2.2.7}$$

Therefore,

$$\zeta = \kappa - \eta. \tag{2.2.8}$$

The diradical character (y) is defined as twice the weight of the doubly excited configuration in the singlet ground state [14]:

$$y \equiv 2\zeta^2 = 1 - 4\kappa\eta. \tag{2.2.9}$$

Here, we use the orthonormal condition $2(\kappa^2 + \eta^2) = 1$ in Eq. (2.1.11a). By substituting Eq. (2.1.13) into Eq. (2.2.9), we obtain [13]

$$y = 1 - \frac{1}{\sqrt{1 + \left(\frac{U}{4t_{ab}}\right)^2}}, \tag{2.2.10}$$

which indicates that the diradical character is a function of U and t_{ab}. By using $r_t(= |t_{ab}/U|)$, y is rewritten as [13]

$$y = 1 - \frac{1}{\sqrt{1 + \left(\frac{1}{4r_t}\right)^2}}. \tag{2.2.11}$$

Figure 2.2 shows the variation of y as a function of $|t_{ab}/U|(\equiv 1/r_t)$. In the case of $|U/t_{ab}| \to \infty(r_t \to 0)$, y value approaches 1, while in the case of $|U/t_{ab}| \leq \sim 1(r_t \geq \sim 1)$, it is close to 0. Because the transfer integral t_{ab} and the effective Coulomb repulsion U represent the ease and difficulty of electron transfer between sites A and B, respectively, $|U/t_{ab}| \to \infty(r_t \to 0)$ implies the localization of electrons on each site, leading to a pure diradical. On the contrary, the delocalization of electrons over two sites is realized for $|U/t_{ab}| \leq \sim 1(r_t \geq \sim 1)$, corresponding to a stable bond. Variation in the mobility of electrons between sites A and B corresponds to that in the diradical character. For example, the elongation of the inter-site distance leads to the decrease in t_{ab} and the increase in U, which causes the decrease in r_t, i.e., the increase in y.

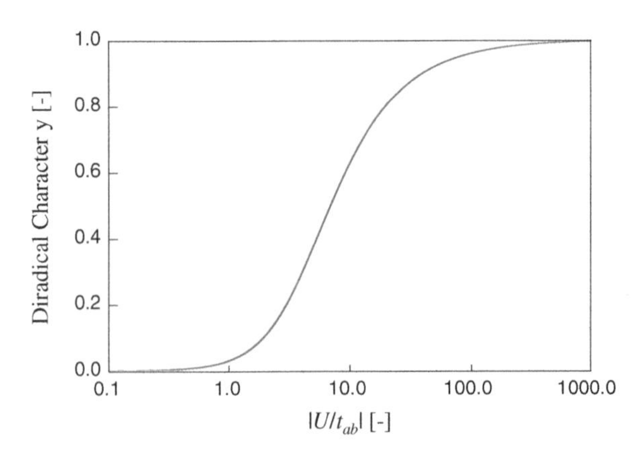

Fig. 2.2 Diradical character y versus $|U/t_{ab}|(=1/r_t)$

2.2.2 Diradical Character in the Spin-Unrestricted Single Determinant Formalism with Spin-Projection Scheme

As mentioned in the previous section, although the diradical character (y) is originally defined in the multi-configuration (MC)-SCF theory by twice the weight of the doubly excited configuration in the singlet ground state [14], the MC-SCF methods are generally difficult to apply to many electron systems due to its huge computational demand. On the other hand, the single determinant scheme is computationally favorable, while the spin-restricted (symmetry-adapted) single determinant approaches like the spin-restricted Hartree-Fock (RHF) cannot be used for obtaining the y value because of its lack of multiply excited configurations. In contrast, the spin-unrestricted [broken-symmetry (BS)] single determinant approaches like the spin-unrestricted HF (UHF) can take account of the multiply excited configurations with low computational cost, which allows us to define the diradical character within this formalism. Nevertheless, the spin-unrestricted methods suffer from the spin contamination [15], which causes the unphysical inclusion of higher spin states like a triplet state in the singlet ground state. This could overshoot y value as compared to that at the MC-SCF level of theory, which gives correct spin states, since the triplet component originating from the spin contamination is a pure diradical state as shown by Eq. (2.1.9). Yamaguchi proposed an approximate spin-projection scheme, which effectively eliminates the spin contamination from the UHF solution, and presented an efficient calculation scheme of diradical character in the approximate spin-projected spin-unrestricted single determinant approaches [10, 11]. We here briefly explain this scheme.

The ground-state UHF wavefunction is written by

$$\Phi = \left|\psi_1^\alpha \psi_1^\beta \cdots \psi_{\mathrm{H}}^\alpha \psi_{\mathrm{H}}^\beta\right\rangle, \tag{2.2.12}$$

where the MOs for the α and β spins (ψ_i^α and ψ_i^β, respectively) satisfy the orthonormal condition for each spin $\left\langle \psi_i^X \middle| \psi_j^X \right\rangle = \delta_{ij} (X = \alpha, \beta)$, and "H" indicates the HOMO. The UHF MOs can be transformed to the corresponding orbitals (χ_i and η_i) with the unitary transformation [15] as

$$\chi_i = \sum_s \psi_s^\alpha U_{si} \quad \text{and} \quad \eta_j = \sum_s \psi_s^\beta V_{sj}, \tag{2.2.13}$$

the orbital overlap of which is diagonal, $T_{ij} = \left\langle \chi_i \middle| \eta_j \right\rangle = T_i \delta_{ij}$. Because the wavefunction is invariant under the unitary transformation, the UHF wavefunction given by Eq. (2.2.13) is rewritten as

$$\Phi = \left|\chi_1 \bar{\eta}_1 \cdots \chi_{\mathrm{H}} \bar{\eta}_{\mathrm{H}}\right\rangle. \tag{2.2.14}$$

The corresponding orbitals are connected with the natural orbitals (NOs) λ_i and μ_i of the UHF wavefunction by

$$\lambda_i = \frac{1}{\sqrt{2(1+T_i)}}(\chi_i + \eta_i) \quad \text{and} \quad \mu_i = \frac{1}{\sqrt{2(1-T_i)}}(\chi_i - \eta_i), \tag{2.2.15}$$

which satisfy the orthonormal condition:

$$\langle \lambda_i \mid \mu_j \rangle = 0 \quad \text{and} \quad \langle \lambda_i \mid \lambda_j \rangle = \langle \mu_i \mid \mu_j \rangle = \delta_{ij}. \tag{2.2.16}$$

From Eqs. (2.2.15) and (2.2.16), the corresponding orbitals χ_i and η_i are given with the NOs λ_i and μ_i as

$$\chi_i = \frac{\sqrt{2(1+T_i)}}{2}\lambda_i + \frac{\sqrt{2(1-T_i)}}{2}\mu_i. \tag{2.2.17}$$

and

$$\eta_i = \frac{\sqrt{2(1+T_i)}}{2}\lambda_i - \frac{\sqrt{2(1-T_i)}}{2}\mu_i. \tag{2.2.18}$$

Because the coefficients satisfy the following relation,

$$\left\{\frac{\sqrt{2(1+T_i)}}{2}\right\}^2 + \left\{\frac{\sqrt{2(1-T)_i}}{2}\right\}^2 = 1, \tag{2.2.19}$$

Equations (2.2.17) and (2.2.18) can be rewritten with a mixing parameter θ_i as,

$$\chi_i = \cos\theta_i\lambda_i + \sin\theta_i\mu_i \tag{2.2.20}$$

and

$$\eta_i = \cos\theta_i\lambda_i - \sin\theta_i\mu_i. \tag{2.2.21}$$

Here θ_i ranges from 0 to $\pi/4$ due to $\cos\theta_i \geq \sin\theta_i \geq 0$ [see Eq. (2.1.3)]. Using these equations, the overlap integral T_i is expressed by

$$T_i = \cos^2\theta_i - \sin^2\theta_i = \cos 2\theta_i, \tag{2.2.22}$$

where the orthonormal condition of NOs [Eq. (2.2.16)] is used.

We here confirm that λ_i and μ_i are the natural orbitals of the UHF wavefunction. Spin-less one-electron reduced density matrix of a N-electron system is defined as

$$\rho(\boldsymbol{r}_1,\boldsymbol{r}'_1) = N\int d\boldsymbol{r}_2\ldots d\boldsymbol{r}_N \Psi(\boldsymbol{r}_1,\boldsymbol{r}_2,\ldots,\boldsymbol{r}_N)\Psi^*(\boldsymbol{r}'_1,\boldsymbol{r}_2,\ldots,\boldsymbol{r}_N). \quad (2.2.23)$$

For the Hartree–Fock wavefunction $\Psi^{\mathrm{HF}} = |\phi_1\ldots\phi_N\rangle$, it is simplified as

$$\rho(\boldsymbol{r}_1,\boldsymbol{r}'_1) = \sum_a \phi_a(\boldsymbol{r}_1)\phi_a^*(\boldsymbol{r}'_1). \quad (2.2.24)$$

Since the wavefunction is invariant under the unitary transformation, $\rho(\boldsymbol{r}_1,\boldsymbol{r}'_1)$ of the UHF wavefunction can be given with the UHF MOs (ψ^α and ψ^β) and the corresponding orbitals (χ_i and η_i) as

$$\begin{aligned}
\rho(\boldsymbol{r}_1,\boldsymbol{r}'_1) &= \sum_i \psi_i^\alpha(\boldsymbol{r}_1)\psi_i^{\alpha*}(\boldsymbol{r}'_1) + \sum_i \psi_i^\beta(\boldsymbol{r}_1)\psi_i^{\beta*}(\boldsymbol{r}'_1) \\
&= \sum_i \chi_i(\boldsymbol{r}_1)\chi_i^*(\boldsymbol{r}'_1) + \sum_i \eta_i(\boldsymbol{r}_1)\eta_i^*(\boldsymbol{r}'_1) \\
&= \sum_i 2\cos^2\theta_i\lambda_i(\boldsymbol{r}_1)\lambda_i^*(\boldsymbol{r}'_1) + \sum_i 2\sin^2\theta_i\mu_i(\boldsymbol{r}_1)\mu_i^*(\boldsymbol{r}'_1) \\
&= \sum_i (1+T_i)\lambda_i(\boldsymbol{r}_1)\lambda_i^*(\boldsymbol{r}'_1) + \sum_i (1-T_i)\mu_i(\boldsymbol{r}_1)\mu_i^*(\boldsymbol{r}'_1),
\end{aligned} \quad (2.2.25)$$

where Eqs. (2.2.20)–(2.2.22) are used. From the last equality, λ_i and μ_i are shown to be natural orbitals of the UHF solution with the occupation numbers of $1+T_i$ and $1-T_i$, respectively. A corresponding orbital pair (χ_i and η_i) constructs a pair of NOs (λ_i and μ_i), where one of the NO pair (λ_i) possesses an occupation number more than 1 ($1+T_i$), while the occupation of the other NO (μ_i) is less than 1 ($1-T_i$). Namely, the total occupation of a NO pair is shown to be always 2 in the single determinant formalism. Also, the bonding and anti-bonding orbitals for the triplet state of this model obtained by the UHF method coincide with the NOs λ_i and μ_i, respectively, due to $T_i = 0$.

For singlet diradical systems with two electrons in the two highest occupied corresponding orbitals (χ_{H} and η_{H}), the wavefunction of the UHF ground state is described by using the corresponding orbitals as

$$\Phi^{\mathrm{UHF}} = |\chi_{\mathrm{H}}\bar{\eta}_{\mathrm{H}}\rangle. \quad (2.2.26)$$

By substituting Eqs. (2.2.20) and (2.2.21) into this equation, Φ^{UHF} is represented with NOs as

$$\begin{aligned}
\Phi^{\mathrm{UHF}} &= \cos^2\theta|\lambda_{\mathrm{H}}\bar{\lambda}_{\mathrm{H}}\rangle - \sin^2\theta|\mu_{\mathrm{L}}\bar{\mu}_{\mathrm{L}}\rangle - \cos\omega\sin\omega\left(|\lambda_{\mathrm{H}}\bar{\mu}_{\mathrm{L}}\rangle - |\mu_{\mathrm{L}}\bar{\lambda}_{\mathrm{H}}\rangle\right) \\
&= \frac{1+T}{2}|\lambda_{\mathrm{H}}\bar{\lambda}_{\mathrm{H}}\rangle - \frac{1-T}{2}|\mu_{\mathrm{L}}\bar{\mu}_{\mathrm{L}}\rangle - \sqrt{\frac{1-T^2}{2}}\left\{\frac{1}{\sqrt{2}}\left(|\lambda_{\mathrm{H}}\bar{\mu}_{\mathrm{L}}\rangle - |\mu_{\mathrm{L}}\bar{\lambda}_{\mathrm{H}}\rangle\right)\right\},
\end{aligned} \quad (2.2.27)$$

where Eq. (2.2.22) ($T \equiv T_{\mathrm{H}}$) is used. The first term in the right-hand side of Eq. (2.2.27) represents the RHF ground configuration, while the second one indicates the doubly excited configuration from the highest occupied NO to the lowest unoccupied NO. Both of these terms represent singlet states. The UHF singlet ground state includes the RHF ground and doubly excited configurations, which enable the UHF method to describe the diradical nature of systems. On the other hand, the third term, composed of singly excited configurations, represents the triplet component, which is the origin of spin contamination in UHF wavefunctions.

As mentioned in Sect. 2.2.1, the diradical character is defined by twice the weight of the doubly excited configuration. We can define the diradical character of the UHF ground state from Eq. (2.2.27) based on this definition. Since the UHF wavefunction suffers from the spin contamination, we need to remove the triplet component to obtain a pure singlet wavefunction, the procedure of which is referred to as the spin-projection [10–12]. For simplicity, Φ^{UHF} is rewritten by

$$\Phi^{\mathrm{UHF}} = C_{\mathrm{I}}\Phi_{\mathrm{G}} + C_{\mathrm{II}}\Phi_{\mathrm{S}} + C_{\mathrm{III}}\Phi_{\mathrm{D}}, \tag{2.2.28}$$

where $C_{\mathrm{I}} = (1+T)/2$, $C_{\mathrm{II}} = -\sqrt{(1-T^2)/2}$, and $C_{\mathrm{III}} = -(1-T)/2$. The first, second and third terms in the right-hand side of Eq. (2.2.28) denote the ground, singly excited and doubly excited configurations, respectively. The removal of the second term (triplet) from the UHF wavefunction with keeping the ratio of the coefficients of the first and third terms ($C_{\mathrm{I}}/C_{\mathrm{III}}$) provides the spin-projected UHF (PUHF) wavefunction,

$$\Phi^{\mathrm{PUHF}} = C'_{\mathrm{I}}\Phi_{\mathrm{G}} + C'_{\mathrm{III}}\Phi_{\mathrm{D}}. \tag{2.2.29}$$

Here, the coefficients satisfy the relation: $C'_{\mathrm{I}}/C'_{\mathrm{III}} = C_{\mathrm{I}}/C_{\mathrm{III}}$. From the orthogonal condition of Φ^{UHF}, we obtain the relation,

$$\frac{C_{\mathrm{I}}^2}{1-C_{\mathrm{II}}^2} + \frac{C_{\mathrm{III}}^2}{1-C_{\mathrm{II}}^2} = 1. \tag{2.2.30}$$

From Eqs. (2.2.29) and (2.2.30), C'_{I} and C'_{III} can be determined by

$$C'_{\mathrm{I}} = \frac{C_{\mathrm{I}}}{\sqrt{1-C_{\mathrm{II}}^2}} \quad \text{and} \quad C'_{\mathrm{III}} = \frac{C_{\mathrm{III}}}{\sqrt{1-C_{\mathrm{II}}^2}}. \tag{2.2.31}$$

By inserting them into Eq. (2.2.29), we obtain

$$\Phi^{\mathrm{PUHF}} = \frac{C_{\mathrm{I}}}{\sqrt{1-C_{\mathrm{II}}^2}}\Phi_{\mathrm{G}} + \frac{C_{\mathrm{III}}}{\sqrt{1-C_{\mathrm{II}}^2}}\Phi_{\mathrm{D}} = \frac{1+T}{\sqrt{2(1+T^2)}}\Phi_{\mathrm{G}} - \frac{1-T}{\sqrt{2(1+T^2)}}\Phi_{\mathrm{D}}. \tag{2.2.32}$$

Therefore, the diradical character in the PUHF formalism is defined by [10–12]

$$y^{\mathrm{PUHF}} = 2C'^{2}_{\mathrm{III}} = 1 - \frac{2T}{1+T^2}, \tag{2.2.33}$$

where the overlap integral T between the corresponding orbitals can be obtained from the occupation number of the LUNO: $n_{\mathrm{LUNO}} = 1 - T$. From the one-electron reduced density matrix of the PUHF wavefunction, the spin-projected occupation numbers of the HONO and LUNO are represented by

$$n^{\mathrm{PUHF}}_{\mathrm{HONO}} = \frac{(1+T)^2}{1+T^2} = \frac{n^2_{\mathrm{HONO}}}{1+T^2} = 2 - y^{\mathrm{PUHF}} \tag{2.2.34a}$$

and

$$n^{\mathrm{PUHF}}_{\mathrm{LUNO}} = \frac{(1-T)^2}{1+T^2} = \frac{n^2_{\mathrm{LUNO}}}{1+T^2} = y^{\mathrm{PUHF}}, \tag{2.2.35b}$$

where $n_{\mathrm{HONO}} = 1 + T$ and $n_{\mathrm{LUNO}} = 1 - T$ are employed [see Eq. (2.2.25)]. Although these equations are defined in singlet diradical systems in relation to their HONO and LUNO, they can be extended to singlet multiradical systems. For a $2n$-radical system, the perfect-pairing type (only considering a doubly excitation from HONO $-$ i to LUNO $+$ i) spin-projected diradical characters and occupation numbers are defined as follows [10–12].

$$y^{\mathrm{PUHF}}_i = 1 - \frac{2T_i}{1+T_i^2}, \tag{2.2.36}$$

and

$$n^{\mathrm{PUHF}}_{\mathrm{HONO}-i} = 2 - y^{\mathrm{PUHF}}_i, \quad \text{and} \quad n^{\mathrm{PUHF}}_{\mathrm{LUNO}+i} = y^{\mathrm{PUHF}}_i, \tag{2.2.37}$$

where T_i is obtained from the occupation number of $\mathrm{LUNO} + i$: $n_{\mathrm{LUNO}+i} = 1 - T_i$.

2.3 Diradical Character Dependences of Excitation Energies and Transition Properties

We here consider only the singlet states for Chap. 4, where (hyper) polarizabilities of singlet molecular systems are discussed. The excitation energy of the triplet state $|T_{1\mathrm{u}}\rangle$ is discussed in Chap. 5, which focuses on singlet fission. In a symmetric two-site diradical model ($\mathrm{A}^\bullet - \mathrm{B}^\bullet$) introduced in Sect. 2.1, there are three singlet

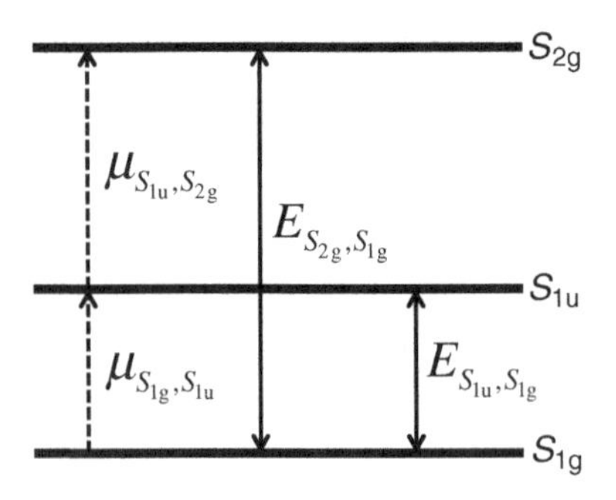

Fig. 2.3 Singlet states (S_{1g}, S_{1u}, S_{2g}) in a two-site diradical model. The excitation energies $\left(E_{S_{1u},S_{1g}}, E_{S_{2g},S_{1g}}\right)$ and transition moments $\left(\mu_{S_{1g},S_{1u}}, \mu_{S_{1u},S_{2g}}\right)$ are shown. The *dashed arrows* indicate optically-allowed transitions

states, $|S_{1g}\rangle$, $|S_{1u}\rangle$ and $|S_{2g}\rangle$. Because $|S_{1g}\rangle$ and $|S_{2g}\rangle$ have the same symmetry (g symmetry), the transition moment between these states disappears. Figure 2.3 shows the three-state model constructed from these singlet states together with the excitation energies and transition moments.

Firstly, we provide the analytical expressions of the excitation energies using several dimensionless physical quantities and diradical characters. Using Eqs. (2.1.10), (2.1.11b), (2.1.12b) and (2.2.11), we obtain [13]

$$\begin{aligned} E_{S_{1u},S_{1g}} &\equiv {}^{1}E_{1u} - {}^{1}E_{1g} = -2K_{ab} + U\left(1 - \frac{1 - \sqrt{1 + 16r_t^2}}{2}\right) \\ &= \frac{U}{2}\left\{1 - 2r_K + \frac{1}{\sqrt{1 - (1 - y)^2}}\right\}, \end{aligned} \tag{2.3.1}$$

and

$$E_{S_{2g},S_{1g}} \equiv {}^{1}E_{2g} - {}^{1}E_{1g} = U\sqrt{1 + 16r_t^2} = \frac{U}{\sqrt{1 - (1 - y)^2}}, \tag{2.3.2}$$

where r_K is defined as $r_K \equiv 2K_{ab}/U$. Second, the analytical expressions of the transition moments along the A–B bond axis are considered. The transition moments between $|S_{1g}\rangle$ and $|S_{1u}\rangle$ are obtained from Eqs. (2.1.10) and (2.1.11a) as [13]

$$
\begin{aligned}
\mu_{S_{1g},S_{1u}} &= -\left\langle S_{1g}\middle|r_1 + r_2\middle|S_{1u}\right\rangle \\
&= -\left\{\kappa\left(\left\langle a\bar{b}\right| + \left\langle b\bar{a}\right|\right) + \eta\left(\left\langle a\bar{a}\right| + \left\langle b\bar{b}\right|\right)\right\}(r_1 + r_2)\left\{\frac{1}{\sqrt{2}}\left(\left|a\bar{a}\right\rangle - \left|b\bar{b}\right\rangle\right)\right\} \\
&= -\frac{1}{\sqrt{2}}\kappa\left\langle\bar{b}\middle|r\middle|\bar{a}\right\rangle + \frac{1}{\sqrt{2}}\kappa\left\langle a\middle|r\middle|b\right\rangle - \frac{1}{\sqrt{2}}\kappa\left\langle b\middle|r\middle|a\right\rangle + \frac{1}{\sqrt{2}}\kappa\left\langle\bar{a}\middle|r\middle|\bar{b}\right\rangle \\
&\quad -\frac{1}{\sqrt{2}}\eta\left\{\left\langle a\middle|r\middle|a\right\rangle + \left\langle\bar{a}\middle|r\middle|\bar{a}\right\rangle\right\} + \frac{1}{\sqrt{2}}\eta\left\{\left\langle b\middle|r\middle|b\right\rangle + \left\langle\bar{b}\middle|r\middle|\bar{b}\right\rangle\right\} \\
&= \sqrt{2}\eta\{(b|r|b) - (a|r|a)\},
\end{aligned}
\tag{2.3.3}
$$

where the following relations are employed:

$$
\begin{cases}
\langle a|r|b\rangle = \langle b|r|a\rangle = \langle\bar{a}|r|\bar{b}\rangle = \langle\bar{b}|r|\bar{a}\rangle = (a|r|b) \\
\langle a|r|a\rangle = \langle\bar{a}|r|\bar{a}\rangle = (a|r|a) \\
\langle b|r|b\rangle = \langle\bar{b}|r|\bar{b}\rangle = (b|r|b)
\end{cases}
. \tag{2.3.4}
$$

Since $(a|r|a)$ and $(b|r|b)$ represent the expectation values of the bond-axis component of the position (r) of electrons using the LNOs a and b, respectively, $(b|r|b) - (a|r|a)$ indicates the dipole $\mu = eR_{BA}$ [with $R_{BA} \equiv R_{bb} - R_{aa} = (b|r|b) - (a|r|a)$, an effective distance between the two radicals], with the e, the electron charge magnitude, equal to 1 in a.u. Thus, we obtain [13]

$$
\mu_{S_{1g},S_{1u}} = \sqrt{2}\eta R_{BA}. \tag{2.3.5}
$$

In the same manner, we obtain the transition moment between $|S_{1u}\rangle$ and $|S_{2g}\rangle$ [13],

$$
\mu_{S_{1u},S_{2g}} = -\langle S_{1u}|r_1 + r_2|S_{2u}\rangle = \sqrt{2}\kappa\{(b|r|b) - (a|r|a)\} = \sqrt{2}\kappa R_{BA}. \tag{2.3.6}
$$

The transition moment ($\mu_{S_{1g},S_{1u}}$) between $|S_{1g}\rangle$ and $|S_{1u}\rangle$ is proportional to the coefficient (η) of the ionic term in the singlet ground state $|S_{1g}\rangle$, while that ($\mu_{S_{1u},S_{2g}}$) between $|S_{1u}\rangle$ and $|S_{2g}\rangle$ is to the coefficient (κ) of the neutral term in $|S_{1g}\rangle$, which is also that of the ionic term in the excited singlet state $|S_{2g}\rangle$. Therefore, the increase in the diradical character (corresponding to the increase in the weight of the neutral term) of the singlet ground state $|S_{1g}\rangle$ leads to the decrease in $\mu_{S_{1g},S_{1u}}$ and the increase in $\mu_{S_{1u},S_{2g}}$. This is because the ionic component of $|S_{1g}\rangle$ ($|S_{2g}\rangle$) decreases (increases) with respect to the increase in the diradical character of $|S_{1g}\rangle$ [see Eqs. (2.1.11a) and (2.1.12a)], while $|S_{1u}\rangle$ keeps the pure ionic nature [see Eq. (2.1.10)]. The transition moment between the ionic ($|a\bar{a}\rangle - |b\bar{b}\rangle$) and neutral ($|a\bar{b}\rangle + |b\bar{a}\rangle$) terms becomes 0, while that between the ionic terms ($|a\bar{a}\rangle - |b\bar{b}\rangle$

and $|a\bar{a}\rangle + |b\bar{b}\rangle)$ has a finite value $(= 2\{(a|r|a) - (b|r|b)\})$. Therefore, larger ionic components in $|S_{1g}\rangle$ and $|S_{2g}\rangle$ leads to larger amplitudes of the transition moments $\mu_{S_{1g},S_{1u}}$ and $\mu_{S_{1u},S_{2g}}$. We here present the expressions of the transition moments as a function of the diradical character y by using Eq. (2.1.14) [13],

$$\begin{aligned}(\mu_{S_{1g},S_{1u}})^2 &= \left(\sqrt{2}\eta R_{\mathrm{BA}}\right)^2 = \frac{8r_t^2 R_{\mathrm{BA}}^2}{\left(1+\sqrt{1+16r_t^2}\right)\sqrt{1+16r_t^2}} \\ &= \frac{R_{\mathrm{BA}}^2}{2}\left\{1-\sqrt{1-(1-y)^2}\right\},\end{aligned} \tag{2.3.7}$$

and

$$\begin{aligned}(\mu_{S_{1u},S_{2g}})^2 &= \left(\sqrt{2}\kappa R_{\mathrm{BA}}\right)^2 = \frac{R_{\mathrm{BA}}^2}{2}\left(1+\frac{1}{\sqrt{1+16r_t^2}}\right) \\ &= \frac{R_{\mathrm{BA}}^2}{2}\left\{1+\sqrt{1-(1-y)^2}\right\}.\end{aligned} \tag{2.3.8}$$

Next, we investigate the diradical character dependences of the transition moments and excitation energies for the singlet three states of the symmetric two-site model, which give us valuable information and help us understand the features of (non)linear optical responses in symmetric diradical systems. Since in usual cases, r_K takes a very small value, we here consider a symmetric two-site system with a fixed R_{BA} and $r_K = 0$. The nondimensional (ND) transition moments and excitation energies are defined by [13]

$$\begin{aligned}\mu_{\mathrm{ND}\ S_{1g},S_{1u}} &\equiv \frac{\mu_{S_{1g},S_{1u}}}{R_{\mathrm{BA}}}, \quad \mu_{\mathrm{ND}\ S_{1u},S_{2g}} \equiv \frac{\mu_{S_{1u},S_{2g}}}{R_{\mathrm{BA}}}, \\ E_{\mathrm{ND}\ S_{1u},S_{1g}} &\equiv \frac{E_{S_{1u},S_{1g}}}{U} \quad \text{and} \quad E_{\mathrm{ND}\ S_{2g},S_{1g}} \equiv \frac{E_{S_{2g},S_{1g}}}{U}\end{aligned} \tag{2.3.9}$$

Figure 2.4 shows the variations of the squared ND transition moments $(\mu_{\mathrm{ND}\ S_{1g},S_{1u}})^2$ and $(\mu_{\mathrm{ND}\ S_{1u},S_{2g}})^2$ with respect to the diradical character y. For $y = 0$, both $(\mu_{\mathrm{ND}\ S_{1g},S_{1u}})^2$ and $(\mu_{\mathrm{ND}\ S_{1u},S_{2g}})^2$ are equal to 0.5. As increasing y from 0 to 1, $(\mu_{\mathrm{ND}\ S_{1g},S_{1u}})^2$ monotonically decreases toward 0, while $(\mu_{\mathrm{ND}\ S_{1u},S_{2g}})^2$ increases toward 1. The diradical character dependences of the ND excitation energies $E_{\mathrm{ND}\ S_{1u},S_{1g}}$ and $E_{\mathrm{ND}\ S_{2g},S_{1g}}$ are shown in Fig. 2.5. With increasing the diradical character, both of the ND excitation energies decrease toward 1, where the decreases are rapid in the small diradical character region, while they are gradual in the intermediate and large diradical character regions. The reduction in the small diradical character region is significant in $E_{\mathrm{ND}\ S_{1u},S_{1g}}$ as compared with $E_{\mathrm{ND}\ S_{2g},S_{1g}}$. Finally, we consider the effect of r_K on the diradical character dependence of $E_{\mathrm{ND}\ S_{1u},S_{1g}}$. From Eqs. (2.3.1), (2.3.2), (2.3.5) and (2.3.6), other

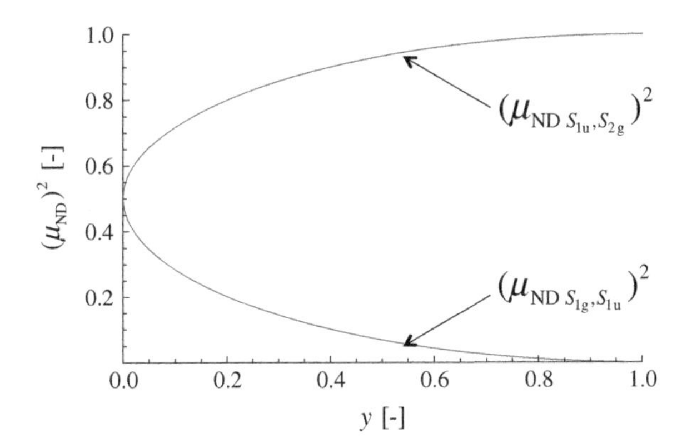

Fig. 2.4 Diradical character (y) dependences of squared nondimensional transition moments $\left(\mu_{ND\ S_{1g},S_{1u}}\right)^2$ and $\left(\mu_{ND\ S_{1u},S_{2g}}\right)^2$

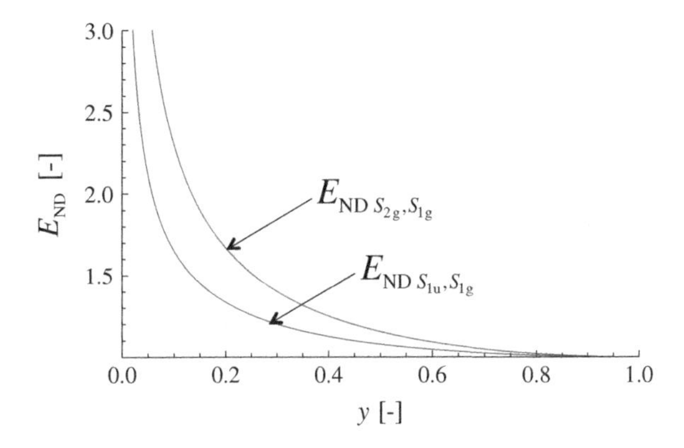

Fig. 2.5 Diradical character (y) dependences of nondimensional excitation energies $E_{ND\ S_{1u},S_{1g}}$ and $E_{ND\ S_{2g},S_{1g}}$ in the case of $r_K = 0$

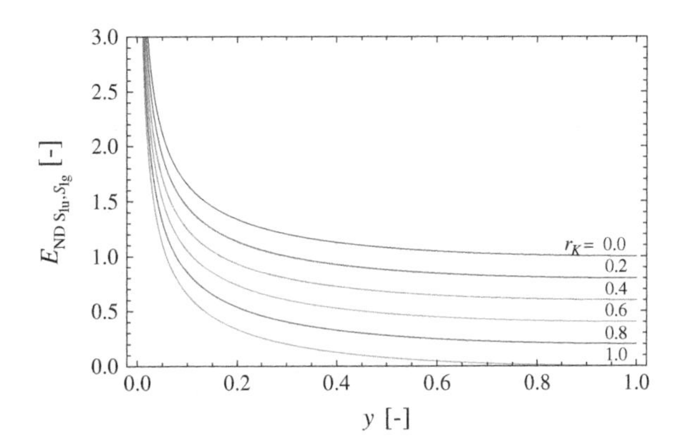

Fig. 2.6 Diradical character (y) dependences of nondimensional excitation energies $E_{ND\ S_{1u},S_{1g}}$ with $r_K = 0.0 - 1.0$

nondimensional singlet excitation energies and transition moments are found to only depend on the diradical character. As seen from Eq. (2.3.1), $E_{ND\ S_{1u},S_{1g}}$ decreases as increasing the diradical character and thus converges to a value, $1 - r_K$, which shows the decrease of $E_{ND\ S_{1u},S_{1g}}$ with the increase of r_K as shown in

Fig. 2.6. As seen from Eq. (2.2.10), the diradical character y tends to increase when U becomes large. It is therefore predicted that the excitation energy $E_{S_{1u},S_{1g}}$ decreases, reaches a stationary value, and in some cases (with very large U) it increases with increasing y values [13, 16]. Such behavior is found to be contrasted with the well-known feature that a closed-shell π-conjugated system exhibits an increase of the oscillator strength of the first optically-allowed excitation and a decrease of the excitation energy with increasing the π-conjugation.

Finally, it is noteworthy that the diradical character y Eq. (2.2.10) can be ex-pressed by the excitation energies [Eqs. (2.1.9), (2.1.10), (2.1.11b), and (2.1.12b)] as

$$y = 1 - \sqrt{1 - \left(\frac{{}^1E_{1u} - {}^3E_{1u}}{{}^1E_{2g} - {}^1E_{1g}}\right)^2} = 1 - \sqrt{1 - \left(\frac{\Delta E_{S(u)} - \Delta E_{T}}{\Delta E_{S(g)}}\right)^2}, \quad (2.3.10)$$

where the first right-hand side includes the energies of the four electronic states (see Sect. 2.1). The second rhs, $\Delta E_{s(g)}(\equiv {}^1E_{2g} - {}^1E_{1g})$, $\Delta E_{s(u)}(\equiv {}^1E_{2u} - {}^1E_{1g})$ and $\Delta E_{T}(\equiv {}^3E_{1u} - {}^1E_{1g})$ correspond to the excitation energies of the higher singlet state of g symmetry (two-photon allowed excited state), of the lower singlet state with u symmetry (one-photon allowed excited state), and of the triplet state with u symmetry, respectively, where $\Delta E_{s(u)}$and $\Delta E_{s(u)}$correspond to the lowest-energy peaks of the one- and two-photon absorption spectra, respectively, while ΔE_T can be obtained from phosphorescence and ESR measurement. Since the diradical character is not an observable but a purely theoretical quantity, this expression is very useful for estimating the diradical character for real molecular systems by experiments [12].

References

1. C. Lambert, Angew. Chem. Int. Ed. **50**, 1756 (2011)
2. Z. Sun, J. Wu, J. Mater. Chem. **22**, 4151 (2012)
3. M. Nishino et al., J. Phys. Chem. A **101**, 705 (1997)
4. M. Nishino et al., Bull. Chem. Soc. Jpn. **71**, 99 (1998)
5. B.O. Roos et al., Angew. Chem. Int. Ed. **46**, 1469 (2007)
6. H. Fukui et al., J. Phys. Chem. Lett. **2**, 2063 (2011)
7. H. Fukui et al., J. Phys. Chem. A **116**, 5501 (2012)
8. C.J. Calzado, J. Cabrero, J.P. Malrieu, R. Caballol, J. Chem. Phys. **116**, 2728 (2002)
9. T. Minami, S. Ito, M. Nakano, J. Phys. Chem. A **117**, 2000 (2013)
10. K. Yamaguchi, Chem. Phys. Lett. **33**, 330 (1975)
11. K. Yamaguchi, in *Self-consistent field: theory and applications*, ed. by R. Carbo, M. Klobukowski (Elsevier, Amsterdam, 1990), p. 727
12. K. Kamada, K. Ohta et al., J. Phys. Chem. Lett. **1**, 937 (2010)
13. M. Nakano, R. Kishi, S. Ohta et al., Phys. Rev. Lett. **99**, 033001 (2007)
14. E.F. Hayes, A.K.Q. Siu, J. Am. Chem. Soc. **93**, 2090 (1971)
15. A.T. Amos, G.G. Hall, Proc. Roy. Soc. **A263**, 483 (1961)
16. R. Kishi, M. Nakano, J. Phys. Chem. A **115**, 3565 (2011)

Chapter 3
Electronic Structures of Asymmetric Diradical Systems

Abstract In this chapter, we present analytical expressions for electronic energies and wavefunctions of the ground and excited states as well as for the excitation energies and transition properties of asymmetric two-site diradical models as the functions of diradical character using the valence configuration interaction method. Such asymmetric diradical systems are realized by symmetric diradical molecules under static electric fields and/or by diradical molecules with asymmetric structures, e.g., donor-acceptor substituted diradicals. Several nondimensional physical factors concerned with "asymmetricity" are introduced in order to describe the electronic structures of these systems.

Keywords Asymmetric diradical system · Diradical character · Excitation energy · Transition moment · Dipole moment difference · Valence configuration interaction

3.1 Asymmetric Diradical Model Using the Valence Configuration Interaction Method

Most of the open-shell singlet molecular systems investigated so far are symmetric molecules except for a few preliminary studies on the static electric field effect on the response properties of symmetric diradical models [1] as well as on the response properties of donor/acceptor-substituted polycyclic aromatic hydrocarbons [2]. Therefore, the effects of an asymmetric electronic distribution on the excitation energies and properties of open-shell molecular systems have not been revealed yet. In this section, we clarify the effects of asymmetric molecular structures [3] and/or of an applied electric field, both of which cause asymmetric electronic distributions.

We present the valence configuration interaction (VCI) scheme based on an asymmetric two-site model $A^\bullet - B^\bullet$ with two electrons in two magnetic orbitals, placed along the bond axis with increasing x. Using the atomic orbitals (AOs) for A and B, i.e., $\chi_A(x)$ and $\chi_B(x)$ [$\chi_A(x) \neq \chi_B(x)$ for asymmetric systems], with

M. Nakano, *Excitation Energies and Properties of Open-Shell Singlet Molecules*,
SpringerBriefs in Electrical and Magnetic Properties of Atoms, Molecules, and Clusters,
DOI 10.1007/978-3-319-08120-5_3

overlap $S_{AB} \equiv \langle \chi_A | \chi_B \rangle$, bonding and anti-bonding molecular orbitals (MOs), $g(x)$ and $u(x)$ can be defined as in the symmetric system:

$$g(x) = \frac{1}{\sqrt{2(1+S_{AB})}}[\chi_A(x) + \chi_B(x)], \quad \text{and}$$
$$u(x) = \frac{1}{\sqrt{2(1-S_{AB})}}[\chi_A(x) - \chi_B(x)], \tag{3.1.1}$$

which are not the canonical MOs of the asymmetric systems when A $\neq$ B. From these MOs, we define the localized natural orbitals (LNOs), $a(x)$ and $b(x)$ [4, 5],

$$a(x) \equiv \frac{1}{\sqrt{2}}[g(x) + u(x)], \quad \text{and}$$
$$b(x) \equiv \frac{1}{\sqrt{2}}[g(x) - u(x)], \tag{3.1.2}$$

which are mainly localized on sites A and B, respectively, similar to $\chi_A(x)$ and $\chi_B(x)$, though they have generally small tails on the other site, satisfying the orthogonal condition, $\langle a|b \rangle = 0$. They become the corresponding AOs [$\chi_A(x)$ and $\chi_B(x)$] at the dissociation limit. The electronic Hamiltonian $\hat{H}$ in atomic units for the asymmetric two-site diradical model is expressed by

$$\hat{H} = -\frac{1}{2}\sum_{i=1}^{N}\nabla_i^2 - \sum_{i=1}^{N}\sum_{X=A}^{B}\frac{Z_X}{r_{iX}} + \sum_{i=1}^{N}\sum_{j>i}^{N}\frac{1}{r_{ij}} = \sum_{i=1}^{N}h(i) + \sum_{i=1}^{N}\sum_{j>i}^{N}\frac{1}{r_{ij}}. \tag{3.1.3}$$

Using the LNOs, the set of electronic configurations is composed of two neutral $\{|a\bar{b}\rangle \equiv |\text{core } a\bar{b}\rangle, |\bar{b}a\rangle \equiv |\text{core } \bar{b}a\rangle\}$ and two ionic $\{|a\bar{a}\rangle \equiv |\text{core } a\bar{a}\rangle, |b\bar{b}\rangle \equiv |\text{core } b\bar{b}\rangle\}$ determinants, where "core" indicates the orthogonal closed-shell core orbitals and the upper-bar (non-bar) indicates the $\beta(\alpha)$ spin. Within that basis, the VCI matrix for zero z-component of spin angular momentum ($M_S = 0$, singlet and triplet) takes the form [3],

$$\begin{pmatrix} \langle a\bar{b}|\hat{H}|a\bar{b}\rangle & \langle a\bar{b}|\hat{H}|b\bar{a}\rangle & \langle a\bar{b}|\hat{H}|a\bar{a}\rangle & \langle a\bar{b}|\hat{H}|b\bar{b}\rangle \\ \langle b\bar{a}|\hat{H}|a\bar{b}\rangle & \langle b\bar{a}|\hat{H}|b\bar{a}\rangle & \langle b\bar{a}|\hat{H}|a\bar{a}\rangle & \langle b\bar{a}|\hat{H}|b\bar{b}\rangle \\ \langle a\bar{a}|\hat{H}|a\bar{b}\rangle & \langle a\bar{a}|\hat{H}|b\bar{a}\rangle & \langle a\bar{a}|\hat{H}|a\bar{a}\rangle & \langle a\bar{a}|\hat{H}|b\bar{b}\rangle \\ \langle b\bar{b}|\hat{H}|a\bar{b}\rangle & \langle b\bar{b}|\hat{H}|b\bar{a}\rangle & \langle b\bar{b}|\hat{H}|a\bar{a}\rangle & \langle b\bar{b}|\hat{H}|b\bar{b}\rangle \end{pmatrix}$$
$$= \begin{pmatrix} 0 & K_{ab} & t_{ab(aa)} & t_{ab(bb)} \\ K_{ab} & 0 & t_{ab(aa)} & t_{ab(bb)} \\ t_{ab(aa)} & t_{ab(aa)} & -h + U_a & K_{ab} \\ t_{ab(bb)} & t_{ab(bb)} & K_{ab} & h + U_b \end{pmatrix}, \tag{3.1.4}$$

where the energy of the neutral determinants $\langle a\bar{b}|\hat{H}|a\bar{b}\rangle (=\langle b\bar{a}|\hat{H}|b\bar{a}\rangle)$ is set to the energy origin. $K_{ab}(\geq 0)$ is a direct exchange integral, and h is the one-electron core Hamiltonian difference, $h \equiv h_{bb} - h_{aa}$, where $h_{pp} \equiv \langle p|h(1)|p\rangle = \langle \bar{p}|h(1)|\bar{p}\rangle \leq 0$ and we here set $h \geq 0$ ($h_{aa} \leq h_{bb}$). Note that since the transfer integrals include, in addition to a one-electron integral (h_{ab}), the two-electron integral between the neutral (e.g., $|a\bar{b}\rangle$) and ionic (e.g., $|a\bar{a}\rangle$) determinants, we obtain two types of transfer integrals, e.g., $t_{ab(aa)} \equiv \langle a\bar{b}|\hat{H}|a\bar{a}\rangle$ and $t_{ab(bb)} \equiv \langle a\bar{b}|\hat{H}|b\bar{b}\rangle$, which are different since $(ab|aa) \neq (ab|bb)$. Subsequently, we introduce the average transfer integral, $t_{ab} \equiv \left(t_{ab(aa)} + t_{ab(bb)}\right)/2$. U_a and U_b are defined by [3]

$$U_a \equiv U_{aa} - U_{ab} + X \quad \text{and} \quad U_b \equiv U_{bb} - U_{ab} - X. \tag{3.1.5}$$

Here, $X \equiv \sum_c^{\text{core}} \{2(U_{ac} - U_{bc}) - (K_{ac} - K_{bc})\}$, which includes the Coulomb (U_{pq}) and exchange (K_{pq}) integrals between LNOs $\{a, b, \text{core}\ (c)\}$, becomes zero for symmetric molecular systems, where $U_a(=U_b)$ indicates the effective Coulomb repulsion defined by the difference between the on-site $[U_{aa}(=U_{bb})]$ and inter-site (U_{ab}) Coulomb integrals. Using the average effective Coulomb repulsion U $[\equiv(U_a + U_b)/2]$, we define the nondimensional quantities [3]:

$$\begin{aligned} &\frac{|t_{ab}|}{U} \equiv r_t(\geq 0), \quad \frac{2K_{ab}}{U} \equiv r_K(\geq 0), \quad \frac{h}{U} \equiv r_h(\geq 0), \\ &\frac{U_a}{U_b} \equiv r_U(\geq 0), \quad \text{and} \quad \left|\frac{t_{ab(aa)}}{t_{ab(bb)}}\right| \equiv r_{tab}(\geq 0). \end{aligned} \tag{3.1.6}$$

The parameter y_S is introduced as an alternative to r_t [3],

$$y_S = 1 - \frac{4r_t}{\sqrt{1 + 16r_t^2}}, \tag{3.1.7}$$

which is the diradical character for the symmetric two-site diradical system with $(r_h, r_U, r_{tab}) = (0, 1, 1)$ [3], but is not for the asymmetric ones, the diradical character of which is referred to as y_A in this case. y_S is thus referred to as “pseudo diradical character” [3], while the diradical character of asymmetric systems is in general a function of $(r_t, r_K, r_h, r_U, r_{tab})$. From Eq. (3.1.6), one observes that the asymmetric electron distribution with more population on A than on B is caused by the increase in $r_h(\geq 0)$ and/or by the decrease in $r_U(\leq 1)$ and/or by $r_{tab}(\leq 1)$. Since the LNOs a and b are well localized in the vicinity of the two atom sites A and B, respectively, we can estimate that the difference between $t_{ab(aa)}$ and $t_{ab(bb)}$ is negligible, i.e., $r_{tab} \sim 1$, as compared to the difference between h_{aa} and h_{bb} and that between U_{aa} and U_{bb}, which are the origin of the asymmetric electron distribution of the present model. We therefore investigate the first cause by changing r_h between 0 and 2 with keeping $(r_U, r_{tab}) = (1, 1)$ for simplicity, which corresponds to the situation where the asymmetricity is primarily governed by the difference of ionization potentials of the constitutive atoms A and B. The VCI eigenvalue equation is thus expressed by

$$\boldsymbol{HC} = \boldsymbol{CE}. \tag{3.1.8}$$

and after using the nondimensional matrix elements we obtain

$$\boldsymbol{H}_{\mathrm{ND}}\boldsymbol{C} = \boldsymbol{CE}_{\mathrm{ND}}, \tag{3.1.9}$$

where $\boldsymbol{E}_{\mathrm{ND}}(\equiv \boldsymbol{E}/U)$ is a nondimensional energy matrix and the explicit form of $\boldsymbol{H}_{\mathrm{ND}}(\equiv \boldsymbol{H}/U)$ is [3]

$$\boldsymbol{H}_{\mathrm{ND}} = \begin{pmatrix} 0 & \frac{r_K}{2} & -\left(\frac{2r_{tab}}{1+r_{tab}}\right)r_t & -\left(\frac{2}{1+r_{tab}}\right)r_t \\ \frac{r_K}{2} & 0 & -\left(\frac{2r_{tab}}{1+r_{tab}}\right)r_t & -\left(\frac{2}{1+r_{tab}}\right)r_t \\ -\left(\frac{2r_{tab}}{1+r_{tab}}\right)r_t & -\left(\frac{2r_{tab}}{1+r_{tab}}\right)r_t & \frac{2r_U}{r_U+1} - r_h & \frac{r_K}{2} \\ -\left(\frac{2}{1+r_{tab}}\right)r_t & -\left(\frac{2}{1+r_{tab}}\right)r_t & \frac{r_K}{2} & \frac{2}{r_U+1} + r_h \end{pmatrix}$$
$$= \begin{pmatrix} 0 & \frac{r_K}{2} & -r_t & -r_t \\ \frac{r_K}{2} & 0 & -r_t & -r_t \\ -r_t & -r_t & 1 - r_h & \frac{r_K}{2} \\ -r_t & -r_t & \frac{r_K}{2} & 1 + r_h \end{pmatrix}. \tag{3.1.10}$$

This indicates that the eigenvalues and eigenvectors depend on the nondimensional quantities (r_t, r_K, r_h, $r_U r_{tab}$) or, considering Eq. (3.1.7), on (y_S, r_K, r_h, r_U, r_{tab}). The eigenvalues and eigenvectors, which are numerically obtained, are represented by $\{j\} = \{\mathrm{T, g, k, f}\}$ (T: triplet state, and g, k, f: singlet states) and $\{C_{a\bar{b},j}, C_{b\bar{a},j}, C_{a\bar{a},j}, C_{b\bar{b},j}\}$, respectively:

$$|\Psi_j\rangle = C_{a\bar{b},j}|a\bar{b}\rangle + C_{b\bar{a},j}|b\bar{a}\rangle + C_{a\bar{a},j}|a\bar{a}\rangle + C_{b\bar{b},j}|b\bar{b}\rangle, \tag{3.1.11}$$

where the $C_{a\bar{b},\mathrm{T}} = -C_{b\bar{a},\mathrm{T}} = 1/\sqrt{2}$ and $C_{a\bar{a},\mathrm{T}} = C_{b\bar{b},\mathrm{T}} = 0$ conditions are satisfied for the triplet state, while $C_{a\bar{b},j} = C_{b\bar{a},j}$ is satisfied for the singlet states [4, 5]. For asymmetric systems, $|C_{a\bar{a},j}| \neq |C_{b\bar{b},j}|$ is satisfied in the case of singlets.

Using the MOs (g and u) [see Eq. (3.1.2)], we can construct an alternative basis set $\{|G\rangle, |S\rangle, |D\rangle\}$ for the singlet states, which represent the ground, the singly-excited and the doubly-excited determinants, respectively [3]:

$$|G\rangle \equiv |g\bar{g}\rangle = \frac{1}{2}\left(|a\bar{a}\rangle + |a\bar{b}\rangle + |b\bar{a}\rangle + |b\bar{b}\rangle\right), \tag{3.1.12a}$$

$$|S\rangle \equiv \frac{1}{\sqrt{2}}(|g\bar{u}\rangle + |u\bar{g}\rangle) = \frac{1}{\sqrt{2}}\left(|a\bar{a}\rangle - |b\bar{b}\rangle\right), \tag{3.1.12b}$$

$$|D\rangle \equiv |u\bar{u}\rangle = \frac{1}{2}\left(|a\bar{a}\rangle - |a\bar{b}\rangle - |b\bar{a}\rangle + |b\bar{b}\rangle\right). \tag{3.1.12c}$$

and the corresponding $\boldsymbol{H}_{\mathrm{ND}}$ is expressed by [3]

$$\begin{aligned}\boldsymbol{H}_{\mathrm{ND}} &= \frac{1}{U}\begin{pmatrix}\langle G|\hat{H}|G\rangle & \langle G|\hat{H}|S\rangle & \langle G|\hat{H}|D\rangle \\ \langle S|\hat{H}|G\rangle & \langle S|\hat{H}|S\rangle & \langle S|\hat{H}|D\rangle \\ \langle D|\hat{H}|G\rangle & \langle D|\hat{H}|S\rangle & \langle D|\hat{H}|D\rangle\end{pmatrix} \\ &= \begin{pmatrix}-2r_t + \frac{1}{2} + \frac{r_K}{2} & \frac{1}{\sqrt{2}}\left(-r_h + \left(\frac{1-r_{tab}}{1+r_{tab}}\right)r_t + \frac{r_U-1}{r_U+1}\right) & \frac{1}{2} \\ \frac{1}{\sqrt{2}}\left(-r_h + \left(\frac{1-r_{tab}}{1+r_{tab}}\right)r_t + \frac{r_U-1}{r_U+1}\right) & 1 - \frac{r_K}{2} & \frac{1}{\sqrt{2}}\left(-r_h - \left(\frac{1-r_{tab}}{1+r_{tab}}\right)r_t + \frac{r_U-1}{r_U+1}\right) \\ \frac{1}{2} & \frac{1}{\sqrt{2}}\left(-r_h - \left(\frac{1-r_{tab}}{1+r_{tab}}\right)r_t + \frac{r_U-1}{r_U+1}\right) & 2r_t + \frac{1}{2} + \frac{r_K}{2}\end{pmatrix} \\ &= \begin{pmatrix}-2r_t + \frac{1}{2} + \frac{r_K}{2} & \frac{-r_h}{\sqrt{2}} & \frac{1}{2} \\ \frac{-r_h}{\sqrt{2}} & 1 - \frac{r_K}{2} & \frac{-r_h}{\sqrt{2}} \\ \frac{1}{2} & \frac{-r_h}{\sqrt{2}} & 2r_t + \frac{1}{2} + \frac{r_K}{2}\end{pmatrix}.\end{aligned} \tag{3.1.13}$$

The last form is for $(r_U, r_{tab}) = (1, 1)$. For symmetric systems, which satisfy $(r_h, r_U, r_{tab}) = (0, 1, 1)$, only $|G\rangle$ and $|D\rangle$ are mixed with each other in the CI calculation due to the disappearance of $\langle G|\hat{H}|S\rangle (= \langle S|\hat{H}|G\rangle = \langle S|\hat{H}|D\rangle = \langle D|\hat{H}|S\rangle)$ [3], while for asymmetric systems ($r_h > 0$, $r_U \neq 1$, $r_{tab} \neq 1$), $|S\rangle$ is mixed with the other determinants due to non-zero $\langle G|\hat{H}|S\rangle$ and $\langle D|\hat{H}|S\rangle$ terms, which originates from the fact that $\{g, u\}$ are not the canonical MOs for $\boldsymbol{H}_{\mathrm{ND}}$. The eigenvectors for the singlet states $\{j\} = \{\mathrm{g, k, f}\}$ are expressed by using the basis set $\{|G\rangle, |S\rangle, |D\rangle\}$. For example, the ground state is represented by

$$|\Psi_{\mathrm{g}}\rangle = \xi|G\rangle + \eta|S\rangle - \zeta|D\rangle, \tag{3.1.14}$$

which satisfies the normalization condition: $\xi^2 + \eta^2 + \zeta^2 = 1$. By comparing Eq. (3.1.11) with (3.1.14), we obtain the relations:

$$\begin{aligned}&\xi = C_{a\bar{b},\mathrm{g}} + \frac{1}{2}\left(C_{a\bar{a},\mathrm{g}} + C_{b\bar{b},\mathrm{g}}\right), \quad \eta = \frac{1}{\sqrt{2}}\left(C_{a\bar{a},\mathrm{g}} - C_{b\bar{b},\mathrm{g}}\right), \text{ and} \\ &\zeta = C_{a\bar{b},\mathrm{g}} - \frac{1}{2}\left(C_{a\bar{a},\mathrm{g}} + C_{b\bar{b},\mathrm{g}}\right).\end{aligned} \tag{3.1.15}$$

Finally, in order to reveal the relationship with the magnetic interaction, we consider the nondimensional energy gap between the singlet (E_{g}) and triplet (E_{T}) ground states. For the symmetric two-site diradical model [$(r_h, r_U, r_{tab}) = (0, 1, 1)$], it is given by [4]

$$r_J = \frac{1}{U}\left(E_{\rm g} - E_{\rm T}\right) = r_K + \frac{1}{2}\left(1 - \sqrt{1 + 16r_t^2}\right), \tag{3.1.16}$$

with $r_J \equiv 2J/U$ and J the effective exchange integral in the Heisenberg Hamiltonian [6]. In this section, we consider the case of $r_K = 0$ (which is satisfied by most common molecular systems with singlet ground state), so that $r_J \leq 0$ is satisfied for the symmetric two-site diradical model as seen from Eq. (3.1.16), i.e., the singlet ground state is equal-lying to or lower-lying than the triplet one. Also, for the symmetric two-site diradical model with $r_K = 0$, when $y_{\rm S}$ tends to 1 [decreasing r_t toward zero value as seen from Eq. (3.1.7)], r_J (negative) tends to 0 [4]. On the other hand, keeping $r_K = 0$, for the asymmetric diradical model, the energy of the triplet state is the same as that of the symmetric diradical model (due to the absence of coupling with the other singlet configurations), while the energy of the singlet ground state is lowered with respect to the symmetric model case due to the non-zero off-diagonal elements (originating from asymmetricity) in Eq. (3.1.13) and therefore, the singlet ground state g is more stable than the triplet ground state T for any $y_{\rm S}$.

3.2 Asymmetricity (r_h) Dependences of the Neutral and Ionic Configurations as well as of the Diradical Character

First, we consider the evolution of the neutral ($P_{\rm N}$) and ionic ($P_{\rm I}$) populations of the three singlet states {g, k, f} as a function of r_h, where $P_{\rm N} = \left|C_{a\bar{b},i}\right|^2 + \left|C_{b\bar{a},i}\right|^2$ and $P_{\rm I} = \left|C_{a\bar{a},i}\right|^2 + \left|C_{b\bar{b},i}\right|^2$ for state i. This is performed for $r_K = 0$ and different values of $y_{\rm S}$(or r_t) (Fig. 3.1). $P_{\rm N}$ and $P_{\rm I}$ display mirror evolution as a function of r_h (with respect to $P_{\rm N} = P_{\rm I} = 1/2$ since $P_{\rm N} + P_{\rm I} = 1$). As seen from Fig. 3.1a and d, for both $y_{\rm S}$ values, $P_{\rm N}$ decreases ($P_{\rm I}$ increases) with r_h in the ground state g, though the difference between $P_{\rm N}$ and $P_{\rm I}$ for $r_h = 0$ is larger for $y_{\rm S} = 0.6$ than for $y_{\rm S} = 0.02$, which reflects the relation $P_{\rm N} = P_{\rm I}$ for $y_{\rm S} = 0$ [3]. More interestingly, the $P_{\rm N}$ and $P_{\rm I}$ curves in the ground state g intersect each other at a finite r_h value, which is larger than 1 for $y_{\rm S} = 0.02$ but for ~ 1 for $y_{\rm S} = 0.6$. On the other hand, in the first excited state k (Fig. 3.1b and e), which is purely ionic for symmetric systems ($r_h = 0$), $P_{\rm I}$ decreases from 1 ($P_{\rm N}$ increases from 0) with r_h, but $P_{\rm I}$ and $P_{\rm N}$ do not intersect each other in the region $0.0 \leq r_h \leq 2.0$, while for $y_{\rm S} = 0.6$, the $P_{\rm I}$ and $P_{\rm N}$ curves cross each other at $r_h \sim 1.0$. This indicates that for intermediate $y_{\rm S}$ values, increasing r_h leads to the inversion of the dominant electronic configurations (neutral/ionic) in states g and k around $r_h = 1$, i.e., the neutral (ionic) configuration is dominant in state g(k) for $r_h < 1$, while the ionic (neutral) is dominant for $r_h > 1$. Such a variation of the dominant configuration in state g explains the decrease of $y_{\rm A}$ toward 0 for $r_h > 1$ as shown in Fig. 3.1d. In contrast to the variations of $P_{\rm N}$ and $P_{\rm I}$ in state k, $P_{\rm I}$ increases ($P_{\rm N}$ decreases) with r_h in state f

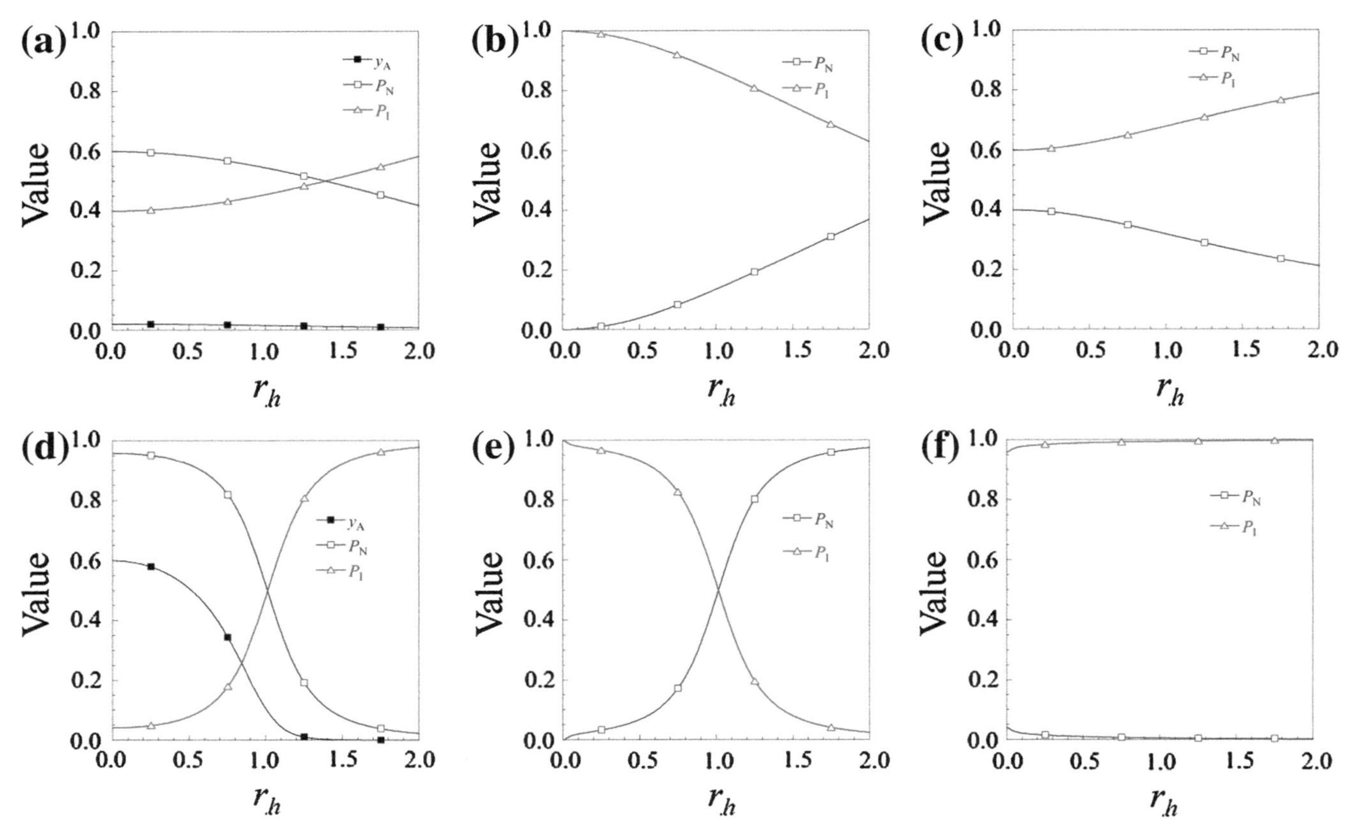

Fig. 3.1 r_h dependences of P_N and P_I in state g, k and f for $y_S = 0.02$ (**a–c**) and 0.6 (**d–f**). Variations in y_A in state g are also shown

(Fig. 3.1c and f), where P_N is larger than P_I at any r_h for $y_S = 0$, though almost pure ionic state is observed at any r_h for $y_S = 0.6$. In order to clarify the origin of the crossing point between P_N and P_I in state g at $r_h \sim 1$ for intermediate y_S, while at $r_h > 1$ for small y_S, we consider the ground state wavefunction with $(r_K, r_U) = (0, 1)$:

$$\left|\Psi_g\right\rangle = \kappa\left(\left|a\bar{b}\right\rangle + \left|b\bar{a}\right\rangle\right) + \eta_1\left|a\bar{a}\right\rangle + \eta_2\left|b\bar{b}\right\rangle \equiv |\mathrm{N}\rangle + |\mathrm{I}\rangle, \tag{3.2.1}$$

where $|\kappa| = |\eta_1| = |\eta_2| = 1/2$ for symmetric systems ($r_h = 0$), while $|\eta_1| > |\eta_2|$ for asymmetric systems with $r_h > 0$. The ground state energy, $\left\langle\Psi_g\middle|\hat{H}\middle|\Psi_g\right\rangle$, can be approximately partitioned into neutral and ionic contributions, referred to as $E_N(=\langle \mathrm{N}|\hat{H}|\mathrm{N}\rangle)$ and $E_I(=\langle \mathrm{I}|\hat{H}|\mathrm{I}\rangle)$, respectively, when $\langle \mathrm{N}|\hat{H}|\mathrm{I}\rangle$ (proportional to transfer integral t_{ab}) is negligible as compared to E_N or E_I, which is reasonable for $y_S > 0$:

$$\begin{aligned} \frac{E_N}{U} &= \kappa^2 K_{ab} \quad \text{and} \\ \frac{E_I}{U} &= \eta_1^2(-h + U_a) + \eta_2^2(h + U_b) + 2\eta_1\eta_2 K_{ab}. \end{aligned} \tag{3.2.2a}$$

Since $P_N = P_I$, which corresponds to the case of $(\kappa^2, \eta_1^2 + \eta_2^2) = (1/4,\ 1/2)$ in Eq. (3.2.1), is approximately satisfied when $E_N = E_I$, we obtain r_h when $P_N = P_I$ from Eq. (3.2.2a) in the case of $r_K = 0$ [3]:

$$r_h = \frac{1}{2}\left(\frac{1}{\eta_1^2 - \eta_2^2}\right), \tag{3.2.2b}$$

which states that for small $y_S(\sim 0)$, corresponding to $\eta_1^2 \sim \eta_2^2$, large $r_h(>1)$ is necessary for achieving $P_N = P_I$, while that for finite $y_S(>0)$, corresponding to $(\eta_1^2, \eta_2^2) \sim (1/2, 0)$, $P_N = P_I$ is achieved at $r_h \sim 1$. This analysis well reproduces the different P_N-P_I intersection behaviors in state g for $y_S = 0.02$ and 0.6 (see Fig. 3.1a and d). The $P_N = P_I$ relationship at $r_h \sim 1$ for $y_S > 0$ is also qualitatively understood by the fact that $r_h \sim 1$, i.e., $h(=h_{bb} - h_{aa}) \sim U$, corresponds to the situation where the attracting and repulsion energies between a pair of electrons in LNOs a and b are similar/identical, which equalizes P_N with P_I in the case of small transfer integral (which tends to be realized for $y_S > 0$).

Next, we consider the asymmetricity (r_h) dependence of the diradical character y_A, which is defined as the occupation number (n_{LUNO}) of the lowest unoccupied natural orbital (LUNO) of state $\left|\Psi_g\right\rangle$ obtained from the diagonalization of its density matrix. Using Eq. (3.1.14), the diradical character y_A is thus expressed as [3]

$$y_A \equiv n_{\mathrm{LUNO}} = 1 - |\xi - \zeta|\sqrt{2 - (\xi - \zeta)^2}, \tag{3.2.3}$$

which is satisfied for both symmetric and asymmetric systems. For symmetric systems, $\eta = 0$, $\xi^2 + \zeta^2 = 1$, and $\xi^2 \geq \zeta^2$ [see Eq. (3.1.14)], so Eq. (3.2.3) reproduces the usual definition that the diradical character is twice the weight of the doubly excitation configuration [7]:

$$y_{\mathrm{A}} = y_{\mathrm{S}} = 2\zeta^2, \tag{3.2.4}$$

Using Eqs. (3.1.15) and (3.2.3), the diradical character y_{A} can also be expressed as a function of the LNO coefficients [3]:

$$y_{\mathrm{A}} = 1 - \left|C_{a\bar{a},\mathrm{g}} + C_{b\bar{b},\mathrm{g}}\right|\sqrt{2 - \left(C_{a\bar{a},\mathrm{g}} + C_{b\bar{b},\mathrm{g}}\right)^2} \tag{3.2.5}$$

where $C_{a\bar{a},\mathrm{g}}$ and $C_{b\bar{b},\mathrm{g}}$ have the same phase in the ground state. For symmetric systems, closed-shell$\left(C_{a\bar{a},\mathrm{g}} = C_{b\bar{b},\mathrm{g}} = C_{a\bar{b},\mathrm{g}} = C_{b\bar{a},\mathrm{g}} = 1/2\right)$ and pure diradical$\left(C_{a\bar{a},\mathrm{g}} = C_{b\bar{b},\mathrm{g}} = 0 \,\text{and}\, C_{a\bar{b},\mathrm{g}} = C_{b\bar{a},\mathrm{g}} = 1/\sqrt{2}\right)$ states [4, 5] correspond to $y_{\mathrm{A}} = 0$ and 1, respectively. On the other hand, for asymmetric systems, this equation indicates that the ionic configuration with asymmetric distribution $\left(\left|C_{a\bar{a},\mathrm{g}}\right| \neq \left|C_{b\bar{b},\mathrm{g}}\right|\right)$ decreases the diradical character y_{A}, e.g., $y_{\mathrm{A}} = 0$ for $\left(C_{a\bar{a},\mathrm{g}} = 1 \,\text{and}\, C_{b\bar{b},\mathrm{g}} = C_{a\bar{b},\mathrm{g}} = C_{b\bar{a},\mathrm{g}} = 0\right)$. Figure 3.2 shows the relationship between the pseudo diradical character y_{S} [Eq. (3.1.7)] and the diradical character y_{A} [Eqs. (3.2.3) and (3.2.5)] for asymmetricity r_h ranging between 0.0 and 2.0. y_{A} is smaller than y_{S}, in particular, in the middle region of y_{S} as increasing the asymmetricity r_h. This is in agreement with the behavior predicted by Eq. (3.2.5) and Fig. 3.1a and d. Interestingly, $y_{\mathrm{S}} = 1$ implies that $y_{\mathrm{A}} = 1$ for $r_h < 1$ but $y_{\mathrm{A}} = 0$ for $r_h > 1$, while y_{A} is close to ~ 0.134 for $r_h = 1$. This behavior originates from the exchange of dominant configuration (neutral/ionic) in state g between $r_h < 1$ and $r_h > 1$ for $y_{\mathrm{S}} > 0$ as shown in Fig. 3.1d. In order to quantitatively clarify this behavior, we consider the eigenvalues and eigenfunctions of Eq. (3.1.9) using Eq. (3.1.13) with (y_{S}, r_K, r_U, r_{tab}) = (1, 0, 1, 1):

(i) For $r_h < 1$,

$$E_{\mathrm{g}} = 0, \quad \left|\Psi_{\mathrm{g}}\right\rangle = \frac{1}{\sqrt{2}}|G\rangle - \frac{1}{\sqrt{2}}|D\rangle = \frac{1}{\sqrt{2}}\left|a\bar{b}\right\rangle + \frac{1}{\sqrt{2}}|b\bar{a}\rangle, \quad y_{\mathrm{A}} = 1, \tag{3.2.6a}$$

$$E_{\mathrm{k}} = 1 - r_h, \quad |\Psi_{\mathrm{k}}\rangle = \frac{1}{2}|G\rangle + \frac{1}{\sqrt{2}}|S\rangle + \frac{1}{2}|D\rangle = |a\bar{a}\rangle, \tag{3.2.6b}$$

$$E_{\mathrm{f}} = 1 + r_h, \quad |\Psi_{\mathrm{k}}\rangle = \frac{1}{2}|G\rangle - \frac{1}{\sqrt{2}}|S\rangle + \frac{1}{2}|D\rangle = \left|b\bar{b}\right\rangle \tag{3.2.6c}$$

(ii) For $r_h = 1$, $\quad E_{\mathrm{g}} = E_{\mathrm{k}}, \quad E_{\mathrm{f}} = 2, \qquad (3.2.6d)$

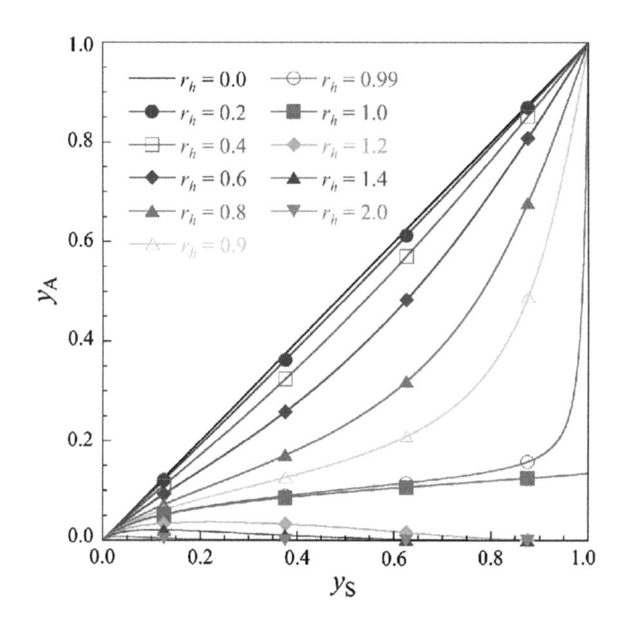

Fig. 3.2 y_S versus y_A plots for $r_h = 0.0 - 2.0$

(iii) For $r_h > 1$,

$$E_g = 1 - r_h, \quad |\Psi_g\rangle = \frac{1}{2}|G\rangle + \frac{1}{\sqrt{2}}|S\rangle + \frac{1}{2}|D\rangle = |a\bar{a}\rangle, \quad y_A = 0, \tag{3.2.6e}$$

$$E_k = 0, \quad |\Psi_k\rangle = \frac{1}{\sqrt{2}}|G\rangle - \frac{1}{\sqrt{2}}|D\rangle = \frac{1}{\sqrt{2}}|a\bar{b}\rangle + \frac{1}{\sqrt{2}}|b\bar{a}\rangle, \tag{3.2.6f}$$

$$E_f = 1 + r_h, \quad |\Psi_k\rangle = \frac{1}{2}|G\rangle - \frac{1}{\sqrt{2}}|S\rangle + \frac{1}{2}|D\rangle = |b\bar{b}\rangle, \tag{3.2.6g}$$

where the y_A values are obtained using Eq. (3.2.5). So, for $r_h < 1$, states g and k are pure neutral (diradical) and ionic while for $r_h > 1$, they are pure ionic and neutral (diradical), respectively. Such inversion of the dominant configurations between states g and k at $r_h = 1$ can also be understood by considering Eq. (3.2.2b) [due to $(\eta_1^2, \eta_2^2) \sim (1/2, 0)$ for non-negligible y_S when $P_N = P_I$] and is exemplified by Fig. 3.1d and e. Namely, in the case of $y_S = 1$, the diradical character y_A is found to abruptly change from 1 to 0 when turning from $r_h < 1$ to $r_h > 1$. Also, when $y_S \to 1$ (y_S is infinitesimally smaller than 1) and $r_h = 1$ ($P_N = P_I$), as shown in Fig. 3.2, y_A asymptotically approaches $1 - \sqrt{3}/2 \sim 0.13397$ since the $(C_{a\bar{a},g}, C_{b\bar{b},g}) \sim (1/\sqrt{2}, 0)$ relationship is satisfied.

3.3 Asymmetricity (r_h) Dependences of the Excitation Energies and Excitation Properties for Different Pseudo Diradical Characters y_S

First, we consider the dependences of the excitation energies and transition moments as a function of y_S for $r_h = 0$, 0.4 and 1.4 (Fig. 3.3a–c). We remind that nondimensional quantities are used, i.e., the energies are divided by the average effective Coulomb repulsion U whereas the transition moments are divided by the dipole $\mu = eR$ [with $R \equiv R_{bb} - R_{aa} = (b|r_1|b) - (a|r_1|a)$, an effective distance between the two radicals], with the e, the electron charge magnitude, equal to 1 in a.u. Also, we assume that $R_{ab} = (a|r_1|b) \sim 0$, since the LNOs a and b are well localized on each site and the overlap between a and b is zero by definition [see Eq. (3.1.2)]. Using the eigenvectors $\{C_{a\bar{b},j}, C_{b\bar{a},j}, C_{a\bar{a},j}, C_{b\bar{b},j}\}$ (j = g, k, f) obtained by solving Eq. (3.1.9), we can evaluate any transition moment μ_{lm} and dipole moment differences $\Delta\mu_{ll}(\equiv \mu_{ll} - \mu_{gg})$ [1, 2].

$$\mu_{lm} \equiv -\langle \Psi_l | \sum_i r_i | \Psi_m \rangle \approx \left(-C_{a\bar{b},l}C_{a\bar{b},m} - C_{b\bar{a},l}C_{b\bar{a},m} - 2C_{b\bar{b},l}C_{b\bar{b},m} \right) R, \quad (3.3.1)$$

and

$$\begin{aligned} \Delta\mu_{ll} &\equiv -\langle \Psi_l | \sum_i r_i | \Psi_l \rangle - \left(-\langle \Psi_g | \sum_i r_i | \Psi_g \rangle \right) \\ &\approx \left[\left(-C^2_{a\bar{b},l} - C^2_{b\bar{a},l} - 2C^2_{b\bar{b},l} \right) - \left(-C^2_{a\bar{b},g} - C^2_{b\bar{a},g} - 2C^2_{b\bar{b},g} \right) \right] R, \end{aligned} \quad (3.3.2)$$

where we used the condition of coordinate independence of these quantities.

For $r_h = 0$ (symmetric system), the diradical character dependences of such quantities were described based on the analytical expressions derived using the VCI two-site diradical model [4]. It is found that (i) states g and f have an equal weight of neutral and ionic configurations, which corresponds to $y_A = y_S = 0$ [see Eqs. (3.2.3)–(3.2.5)], while state k is purely ionic, (ii) only states g and f are correlated, so that the relative increase in the neutral configuration in state g (increase in y_S) is associated with an increase in the ionic configuration in state f, while (iii) state k remains purely ionic for all values of y_S. These changes in the neutral and ionic configurations are at the origin of the behavior of the excitation energies and excitation properties as a function of y_S. As shown in Fig. 3.3a, the excitation energies of states f and k, i.e., E_{fg}/U and E_{kg}/U, where $E_{ij} \equiv E_i - E_j$, decrease with y_S and coincide with each other at $y_S = 1$, while the amplitudes of the first and second transition moments, i.e., $|\mu_{gk}|/R$ and $|\mu_{kf}|/R$, decrease toward zero and increase toward 1, respectively. The $|\mu_{gf}|/R$ remains zero for symmetric systems at any y_S, while for asymmetric systems ($r_h > 0$), the dipole moment differences, $\Delta\mu_{kk}/R$ and $\Delta\mu_{ff}/R$ appear. It is found from Fig. 3.3b that upon increasing y_S, both the amplitudes of $\Delta\mu_{kk}/R$ (positive) and $\Delta\mu_{ff}/R$ (negative)

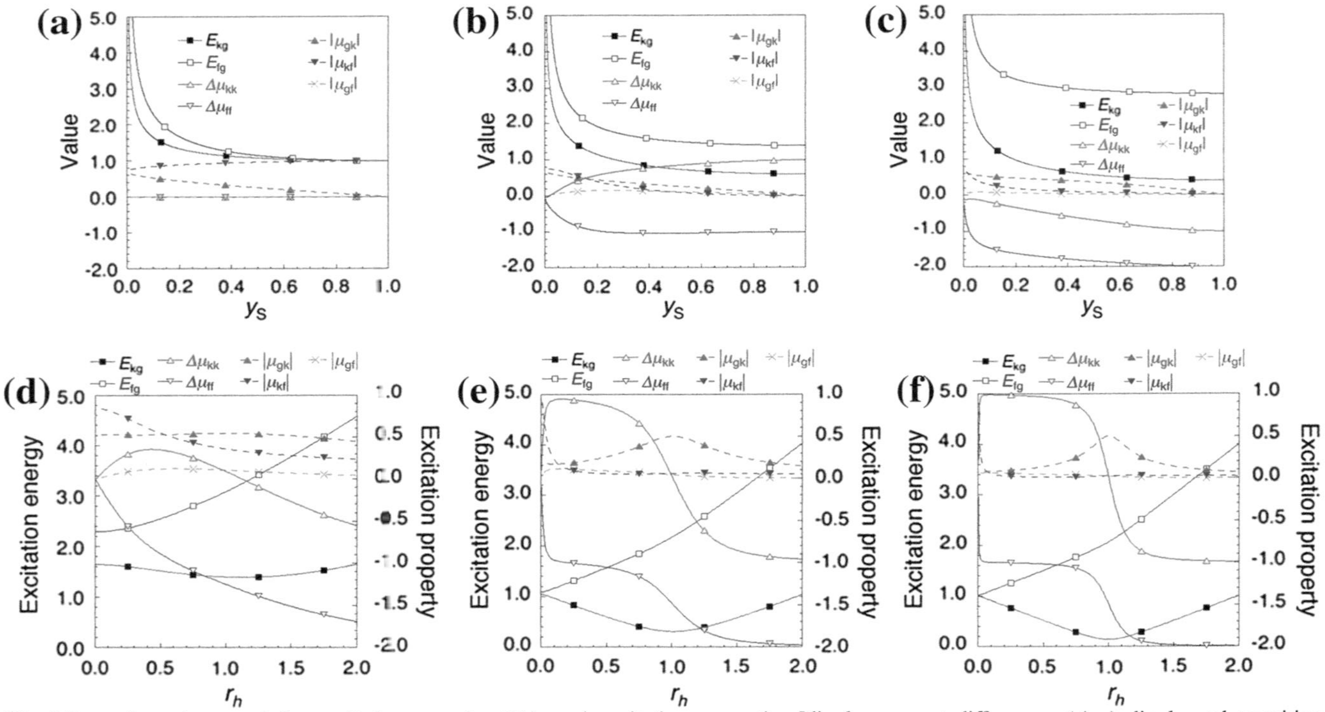

Fig. 3.3 y_S dependences of the excitation energies (E_{ij}), and excitation properties [dipole moment differences ($\Delta\mu_{ii}$) dipole and transition moment amplitudes ($|\mu_{ij}|$)] for $r_h = 0$ (**a**), 0.4 (**b**), and 1.4 (**c**) as well as r_h dependences of those quantities for $y_S = 0.1$ (**d**), 0.6 (**e**), and 0.8 (**f**). Note that nondimensional excitation energies and properties are plotted

increase. These dipole moment differences indicate that the polarizations in states g and k have the same direction (from A to B), which is opposite to that in state f (see, for example, wavefunction distributions in Fig. 3.4c). Then, both $|\mu_{gk}|/R$ and $|\mu_{kf}|/R$ decrease toward zero as increasing y_S by contrast to the symmetric case ($r_h = 0$) (Fig. 3.3a). It is also found that both E_{fg}/U and E_{kg}/U decrease as a function of y_S, while, for the whole y_S range, they are larger and smaller than those of the symmetric case ($r_h = 0$), respectively. This is also understood by the relations: $E_{fg}/U = 1 + r_h$ and $E_{kg}/U = 1 - r_h$ at $y_S = 1$ [see Eqs. (3.2.6a)–(3.2.6c)]. Furthermore, when r_h becomes larger than 1, the dominant configurations are exchanged between states g and k, which leads to the sign change of $\Delta\mu_{kk}/R$ at $r_h \sim 1$ ($\Delta\mu_{kk}/R > 0$ for $r_h < 1$ and $\Delta\mu_{kk}/R < 0$ for $r_h > 1$), and for $r_h > 1$, its amplitude increases with y_S due to the increase of diradical (ionic) configuration in state k(g) as shown in Fig. 3.4c. It is also found from Eqs. (3.2.6e)–(3.2.6g) that $E_{fg}/U = 2r_h$ and $E_{kg}/U = = r_h - 1$ at $y_S = 1$ for $r_h > 1$, which causes the increase in E_{kg}/U after reaching a minimum at $r_h \sim 1$ though $E_{fg}/U = 2r_h$ continues to increase with r_h as shown in Fig. 3.4c. On the other hand, for $r_h \sim 1$, E_{kg}/U reduces to 0 as increasing y_S, while $\Delta\mu_{kk}/R$ is close to 0 and $|\mu_{gk}|/R$ is almost constant (and non-zero) in the whole y_S range. This reflects the equal weight of the neutral and ionic configurations in both states as shown in Fig. 3.1d

Next, we investigate the r_h dependences of the excitation energies and excitation properties for different pseudo diradical characters $y_S = 0.1$, 0.6 and 0.8 (see Figs. 3.3d–f). In case of small pseudo diradical character ($y_S = 0.1$), E_{kg}/U remains almost constant in the whole region of r_h, while E_{fg}/U increases with r_h. Moreover, $|\mu_{gk}|/R$ ($|\mu_{gf}|/R$) remains $\sim$0.5 (negligible), while $|\mu_{kf}|/R$ decreases with r_h. The increase followed by a decrease of $\Delta\mu_{kk}/R$ as a function of r_h, which is positive then negative for $r_h < 1$ and $r_h > 1$, respectively, reflect the facts: (i) the increase of ionic configuration in state g and the increase of the asymmetricity of the ionic configuration in state k with $r_h(<1)$, (ii) the inversion of the dominant configurations (neutral/ionic) in states g and k at $r_h \sim 1$, (iii) the increase of the dominant ionic and neutral configurations in states g and k, respectively, with $r_h(>1)$, and (iv) the dominant asymmetric ionic configuration in state f (with polarization in the opposite direction to those in states g and k) rapidly grows with the slight increase in r_h. Then, the relative increase of the ionic configuration with asymmetric distribution in state g, the relative decrease in the neutral configuration in state k particularly for $r_h > 1$, and the weight of nearly pure ionic configuration (with the opposite polarization direction to that of states g and k) in state f are found to be more augmented with y_S. These variations are also shown in those of P_N and P_I in each state (see Fig. 3.1d–f). In particular, the rapid change of $\Delta\mu_{kk}/R$ from positive to negative value (with its significant change in amplitude), the maximization of $|\mu_{gk}|/R$, and the minimization of E_{kg}/U at $r_h \sim 1$ [see Eqs. (3.2.6a)–(3.2.6g)] represent characteristic r_h-dependent behaviors of the excitation energies and excitation properties in asymmetric diradical singlet systems, which are predicted to govern the r_h dependences of β and γ [3].

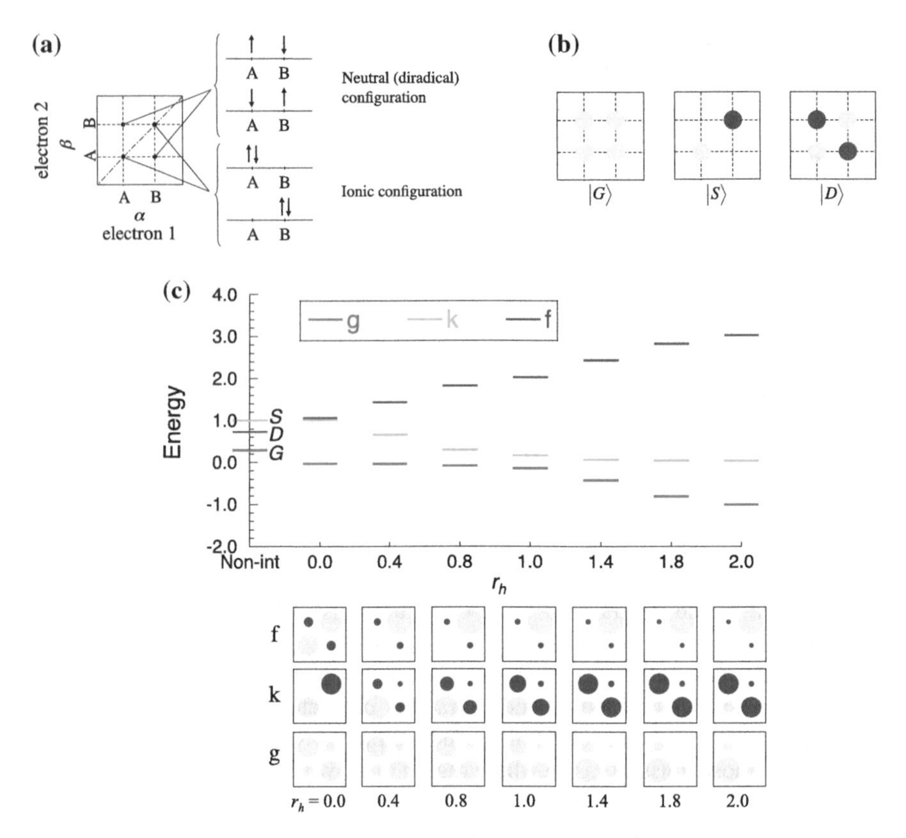

Fig. 3.4 Schematic representation of the spatial wavefunction on the plane (1a, 2b) for the one-dimensional two-electron molecular system (**a**). The schematic wavefunctions of $|G\rangle,|S\rangle$, and $|D\rangle$ are shown in (**b**). Variations of energies and schematic wavefunctions of three states {g, k, f} for $y_S = 0.6$ are shown in (**c**). Energy states of $|G\rangle,|S\rangle$, and $|D\rangle$ are shown by *red, green, blue bars*, respectively, at "Non-int". *Yellow* and *blue circles* represent the positive and negative phases of wavefunctions, respectively, while the size of the circles determines the amplitudes of the wavefunction coefficients

Finally, for further understanding the above r_h dependences of the excitation energies and excitation properties, we provide schematic diagrams of the spatial distributions of the wavefunctions for state {g, k, f} [1, 3, 8] As shown in Fig. 3.4a, the dotted lines indicate the positions of nuclei A and B, and the diagonal dashed line represents the Coulomb wall. The black circles symmetrically distributed with respect to the diagonal dashed line represent the neutral (covalent or diradical) configurations, while those on the diagonal dashed line represents the ionic configurations. So, the wavefunctions, $|G\rangle,|S\rangle$, and $|D\rangle$ [Eq. (3.1.12a)–(3.1.12c)], are illustrated by using the spatial distribution on the $(1\alpha, 2\beta)$ plane see Fig. 3.4b), where yellow and blue circles represent the positive and negative phases of the wavefunctions, respectively. Figure 3.4c sketches the evolution of the state energies of {g, k, f} with $y_S = 0.6$ for $r_h = 0.0 - 2.0$ as well as for

$|G\rangle$,$|S\rangle$, and $|D\rangle$. For $r_h = 0.0$ (symmetric system), state k and g are close to each other due to the electron correlation between states $|G\rangle$ and $|D\rangle$. As explained in Fig. 3.1d–f, for $r_h = 0$, dominant neutral configurations are observed in state g, while pure and dominant ionic configurations for states k and g, respectively, with $y_S = 0.6$ (see $r_h = 0$ in Fig. 3.4c). Then, as increasing r_h, the energy intervals, g-f(E_{fg}) and k-f(E_{fk}) grow, while E_{kg} decreases up to $r_h \sim 1$ and increases beyond, which results in a minimum E_{kg} at $r_h \sim 1$. Again, this is associated with the exchange of dominant configurations between states g and k at $r_h \sim 1$. Indeed, as seen from Fig. 3.4c, the increase in r_h leads to an asymmetric distribution on (A, A) (see Fig. 3.4a) for states g and k, as well as to a relative increase of the ionic and diradical configurations in states g and k, respectively. As a result, at $r_h \sim 1$, the spatial distributions of the wavefunctions are nearly the same for states g and k except for the phases, the feature of which causes the rapid decrease of $\Delta\mu_{kk}/R$ toward zero as well as the local maximum of $|\mu_{gk}|/R$ as shown in Fig. 3.3e. The same trend, i.e., relative increase of the diradical and ionic configurations in states k and g, respectively, remains after $r_h \sim 1$, so that states g and k approach asymmetric pure ionic and pure diradical configurations, respectively, at large r_h (see $r_h = 2$ in Fig. 3.4c). In contrast, as seen from Fig. 3.4c, the spatial wavefunction distribution of state f retains an almost pure ionic character with opposite polarization direction to those of states g and k, i.e., localized on (B, B), in a wide range of r_h, which accounts for the strong decrease of $|\mu_{gf}|/R$ and $|\mu_{kf}|/R$ and the enhancement of $|\Delta\mu_{ff}|/R$ in a wide range of r_h as shown in Fig. 3.3e. It is also noted that, owing to the choice of $h(\equiv h_{bb} - h_{aa}) > 0$, an increase in population at (A, A) [(B, B)] tends to stabilize [destabilize] the state, which explains the energy evolution for states {g, k, f} as a function of r_h as shown in Fig. 3.4c.

References

1. M. Nakano, B. Champagne et al., J. Chem. Phys. **133**, 154302 (2010)
2. M. Nakano, T. Minami et al., J. Phys. Chem. Lett. **2**, 1094 (2011)
3. M. Nakano, B. Champagne, J. Chem. Phys. **138**, 244306 (2013)
4. M. Nakano et al., Phys. Rev. Lett. **99**, 033001 (2007)
5. C.J. Calzado et al., J. Chem. Phys. **116**, 2728 (2002)
6. W. Heisenberg, Z. Phys. **49**, 619 (1928)
7. E.F. Hayes, A.K.Q. Siu, J. Am. Chem. Soc. **93**, 2090 (1971)
8. M. Nakano, R. Kishi et al., J. Chem. Phys. **125**, 074113 (2006)

Chapter 4
Diradical Character View of (Non)Linear Optical Properties

Abstract In this chapter, we clarify the diradical character dependences of (hyper)polarizabilities, which are molecular origins of (non)linear optical responses, in static and resonant cases based on the diradical character dependences of excitation energies and properties of two-site symmetric and asymmetric diradical models with two electrons in two active orbitals. The analysis results highlight the differences of optical response properties between open-shell and closed-shell molecular systems, and contribute to the construction of novel molecular design guidelines for highly efficient nonlinear optical systems.

Keywords Nonlinear optics · Hyperpolarizability · Diradical character · Excitation energy · Transition moment · Molecular design

4.1 (Non)Linear Optical Properties: (Hyper)Polarizabilities

First, we briefly outline the perturbative descriptions of (non)linear optical (NLO) susceptibilities, i.e., (hyper)polarizabilities at the molecular scale, for later sections. When a material is irradiated with light (described by external electric field $\boldsymbol{F}'$), macroscopic polarization $\boldsymbol{P}$ is induced by the applied electric field $\boldsymbol{F}'$. The i-axis component of the polarization is given by a power series of $\boldsymbol{F}'$,

$$P^i = \sum_j \chi_{ij}^{(1)} F'^j + \sum_{jk} \chi_{ijk}^{(2)} F'^j F'^k + \sum_{jkl} \chi_{ijkl}^{(3)} F'^j F'^k F'^l + \cdots . \tag{4.1.1}$$

Here, $\chi_{ij}^{(1)}$, $\chi_{ijk}^{(2)}$ and $\chi_{ijkl}^{(3)}$ represent the linear optical susceptibility, and the second- and third-order NLO susceptibilities, respectively, and i, j, k and l indicate the components of the orthogonal coordinate axes (x, y, z). In the usual case with small $\boldsymbol{F}'$ amplitudes, the polarization can be described by the first term of the right-hand side of Eq. (4.1.1), while in the presence of intense light such as the laser field, the higher-order terms are no longer negligible. These terms cause various types of NLO polarizations in substances, which are the origins of NLO phenomena such

M. Nakano, *Excitation Energies and Properties of Open-Shell Singlet Molecules*, SpringerBriefs in Electrical and Magnetic Properties of Atoms, Molecules, and Clusters, DOI 10.1007/978-3-319-08120-5_4

as high harmonic generations and intensity-dependent refractive index. The macroscopic polarization originates from the polarization at the molecular level (microscopic polarization). Similar to the case of the macroscopic polarization, the microscopic polarization $\boldsymbol{p}$ is expanded by the power series of electric field $\boldsymbol{F}$,

$$\begin{aligned} p^i &= \sum_j \alpha_{ij} F^j + \sum_{jk} \beta_{ijk} F^j F^k + \sum_{jkl} \gamma_{ijkl} F^j F^k F^l + \cdots \\ &= p^{(1),i} + p^{(2),i} + p^{(3),i} + \cdots, \end{aligned} \tag{4.1.2}$$

where F^i represents the i-axis component of the local electric field at the molecular site, while the coefficients α_{ij}, β_{ijk} and γ_{ijkl} indicate the polarizability, first hyperpolarizability and second hyperpolarizability, respectively. Because these properties strongly depend on the structure and electronic state of a molecule, the fundamental studies exploring the structure–property relationships at the molecular level are important for understanding the NLO phenomena as well as for designing NLO materials.

The perturbation approach, i.e., summation-over-states (SOS) approach, is a strong tool for analysis of the first and second hyperpolarizabilities because this approach connects these properties with the transition energies, transition moments and dipole moment differences between states, which can be obtained by the quantum chemical calculations and be also inferred from the molecular orbitals and their energies. The SOS expressions of α, β and γ derived from a perturbation theory expansion of the energy are written as follows [1–5],

$$\alpha_{ij}(-\omega_1;\omega_1) = \frac{1}{\hbar}\sum_{n\neq 0}\left(\frac{\mu_{0n}^i \mu_{n0}^j}{\omega_{n0}+\omega_1} + \frac{\mu_{0n}^i \mu_{n0}^j}{\omega_{n0}-\omega_1}\right), \tag{4.1.3}$$

$$\begin{aligned} \beta_{ijk}(-(\omega_1+\omega_2);\omega_1,\omega_2) &= \frac{1}{2\hbar^2} P(i,j,k;-(\omega_1+\omega_2),\omega_1,\omega_2) \\ &\quad \times \sum_{n\neq 0}\sum_{m\neq 0} \frac{\mu_{0n}^i \bar{\mu}_{nm}^j \mu_{m0}^k}{\{\omega_{0n}-(\omega_1+\omega_2)\}(\omega_{0n}-\omega_2)}, \end{aligned} \tag{4.1.4}$$

and

$$\begin{aligned} \gamma_{ijkl}(-(\omega_1+\omega_2+\omega_3);\omega_1,\omega_2,\omega_3) &= \frac{1}{6\hbar^3} P(i,j,k,l;-(\omega_1+\omega_2+\omega_3);\omega_1,\omega_2,\omega_3) \\ &\quad \times \left[\sum_{n\neq 0}\sum_{m\neq 0}\sum_{n'\neq 0} \frac{\mu_{0n}^i \bar{\mu}_{nm}^j \bar{\mu}_{mn'}^k \mu_{n'0}^l}{\{\omega_{n0}-(\omega_1+\omega_2+\omega_3)\}(\omega_{m0}-\omega_2-\omega_3)(\omega_{n'0}-\omega_3)} \right. \\ &\quad \left. - \sum_{n\neq 0}\sum_{m\neq 0} \frac{\mu_{0n}^i \mu_{n0}^j \mu_{0m}^k \mu_{m0}^l}{\{\omega_{n0}-(\omega_1+\omega_2+\omega_3)\}(\omega_{m0}-\omega_2-\omega_3)(\omega_{m0}-\omega_3)}\right], \end{aligned} \tag{4.1.5}$$

where subscript "0" represents the ground state, ω_{n0} is defined by the energy difference between the ground (0) and excited state (n) as $E_n - E_0 \equiv \hbar\omega_{n0}$, and $\bar{\mu}^j_{nm} \equiv \mu^j_{nm} - \mu^j_{00}\delta_{nm}$ indicates the transition moment ($n \neq m$) or dipole moment difference ($n = m$). $P(i,j,k; -(\omega_1 + \omega_2), \omega_1, \omega_2)$ is a permutation operator, which implies that for any permutation of (i,j,k), an equivalent permutation of $(-(\omega_1 + \omega_2), \omega_1, \omega_2)$ is made simultaneously.

As an example, the SOS expression of γ [Eq. (4.1.5)] is explained in static case ($\omega_1 = \omega_2 = \omega_3 = 0$). For NLO materials for frequency conversion, dynamic γ in off-resonant frequency region is important and such γ is approximately described by the static γ. Here, we consider a system with a large diagonal component of γ (γ_{iiii}), which characterizes the third-order optical nonlinearity of the system. The diagonal component of static γ is obtained by inserting $\omega_1 = \omega_2 = \omega_3 = 0$ into Eq. (4.1.5):

$$\gamma_{iiii}(0) = 4\sum_{n\neq 0}\frac{(\mu^i_{n0})^2(\Delta\mu^i_{nn})^2}{(E_{n0})^3} - 4\sum_{n\neq 0, m\neq 0}\frac{(\mu^i_{n0})^2(\mu^i_{m0})^2}{(E_{n0})^2E_{m0}}$$
$$+ 8\sum_{n\neq 0, m\neq 0, n\neq m}\frac{\mu^i_{0n}\Delta\mu^i_{nn}\mu^i_{nm}\mu^i_{m0}}{(E_{n0})^2E_{m0}} + 4\sum_{\substack{n\neq 0, m\neq 0, n'\neq 0\\ n\neq m, m\neq n'}}\frac{\mu^i_{0n}\mu^i_{nm}\mu^i_{mn'}\mu^i_{n'0}}{E_{n0}E_{m0}E_{n'0}}. \quad (4.1.6)$$

where we use $E_n - E_0 = \hbar\omega_{n0}$ and the definition of $\bar{\mu}^i_{nm}$:

$$\bar{\mu}^i_{nm} = \mu^i_{nm} - \mu^i_{00}\delta_{nm} = \begin{cases} \mu^i_{nn} - \mu^i_{00} \equiv \Delta\mu^i_{nn} \ (n = m), \\ \mu^i_{nm} \ (n \neq m) \end{cases} \quad (4.1.7)$$

The right-hand side terms of Eq. (4.1.6) are classified into three types of virtual excitation processes [6–8]. The first term (type I) represents the process (0–n–n–n–0) (see Fig. 4.1a), which includes the transition moments (μ^i_{n0}), dipole moment differences between the ground (0) and the nth excitation states ($\Delta\mu^i_{nn}$), and excitation energies (E_{n0}). This term vanishes in centrosymmetric systems due to $\Delta\mu^i_{nn} = 0$. The second term (type II) represents the process (0–n–0–m–0) (Fig. 4.1a), which involves symmetric ($n = m$) and asymmetric ($n \neq m$) processes. Contrary to other terms, type II has negative contributions to γ, which implies that the relative enhancement of type II could lead to negative γ. The third term (type III-1) consists of two processes (0–n–n–m–0) and (0–m–n–n–0) originating in the conditions ($n = m$ and $m \neq n'$) and ($n \neq m$ and $m = n'$), respectively, which results in the factor of 2 in type III-1 (see Fig. 4.1a). It is noteworthy that the two conditions lead to two different terms for dynamic γ. The type III-1 also disappears in centrosymmetric systems due to $\Delta\mu^i_{nn} = 0$. The last term (type III-2) represents the (0–n–m–n'–0) process (Fig. 4.1a) involving symmetric ($n = n'$) and asymmetric ($n \neq n'$) processes as in the case of type I. The type III-1 and type III-2 possess the transition moments between excited states. Equation (4.1.6) indicates that larger transition moments and/or dipole moment differences, and/or smaller

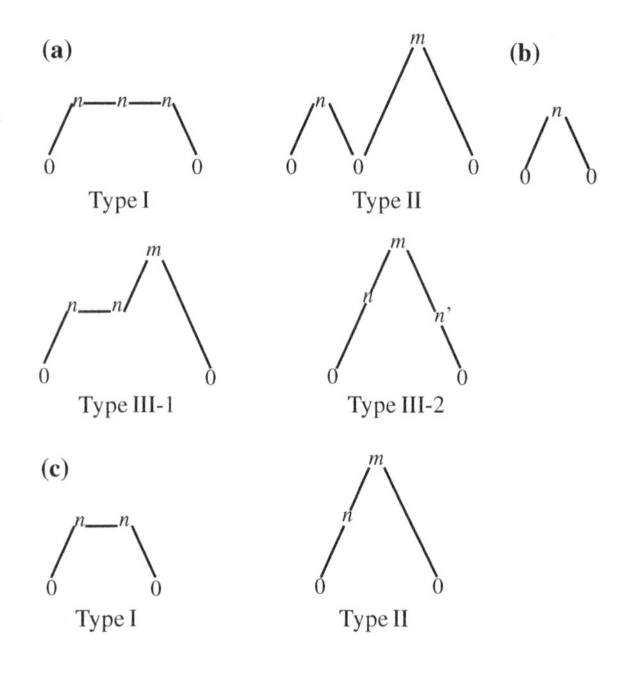

Fig. 4.1 Virtual excitation processes of static γ (**a**), α (**b**) and β (**c**) in their SOS expressions. Symbols 0, n, m, n' indicate the ground, nth, mth and n'th excited states, respectively

excitation energies could lead to large static γ. On the other hand, the component (type II term) in Eq. (4.1.6) is related with polarizability (α_{ii}) because the SOS expression of static α_{ii} consists of μ^i_{n0} and E_{n0}:

$$\alpha_{ii}(0) = 2\sum_{n \neq 0} \frac{(\mu^i_{0n})^2}{E_{n0}}, \tag{4.1.7}$$

which is shown in Fig. 4.1b. Therefore, if a system has a very large α_{ii}, the type II term of γ_{iiii} also has a very large amplitude, which can lead to negative γ_{iiii}. In summary, construction of guiding principles based on these parameters is indispensable for efficient molecular designs of third-order NLO compounds.

In a similar manner, the static β for asymmetric systems is obtained by

$$\beta_{iii}(0) = 3\sum_{n(\neq 0)} \frac{(\mu^i_{0n})^2 \Delta\mu^i_{nn}}{(E_{n0})^2} + 3\sum_{n\neq 0, m\neq 0, n\neq m} \frac{\mu^i_{0n}\mu^i_{nm}\mu^i_{m0}}{E_{n0}E_{m0}}, \tag{4.1.8}$$

where the first and second terms in the right-hand side are referred to as type I and II terms (see Fig. 4.1c). These terms are found to inverse the sign if the coordinate axis is inverted, the fact of which implies that the β_{iii} of centrosymmetric systems (with respect to the i-axis) vanish.

4.2 Second Hyperpolarizability (γ) for Symmetric Systems

For symmetric systems (with respect to the spin polarization direction), the type I and III-1 terms, i.e., the first and the third terms in Eq. (4.1.6), respectively, disappear due to $\Delta\mu_{nn}^{i} = 0$ and reduce to the following equation, which implies that the SOS expression of γ_{iiii} for symmetric systems is described by type II (γ^{II}) and type III-2 terms ($\gamma^{\text{III-2}}$) [9],

$$\gamma_{iiii}(0) = \gamma_{iiii}^{\mathrm{II}} + \gamma_{iiii}^{\mathrm{III-2}} = -4 \sum_{n \neq 0, m \neq 0} \frac{(\mu_{n0}^{i})^2 (\mu_{m0}^{i})^2}{(E_{n0})^2 E_{m0}} + 4 \sum_{\substack{n \neq 0, m \neq 0, n' \neq 0 \\ n \neq m, m \neq n'}} \frac{\mu_{0n}^{i} \mu_{nm}^{i} \mu_{mn'}^{i} \mu_{n'0}^{i}}{E_{n0} E_{m0} E_{n'0}}. \quad (4.2.1)$$

In a symmetric two-site diradical model ($\mathrm{A}^{\bullet} - \mathrm{B}^{\bullet}$) introduced in Sect. 2.1, there are three singlet states, $|S_{1g}\rangle$, $|S_{1u}\rangle$ and $|S_{2g}\rangle$. The transition moment between $|S_{1g}\rangle$ and $|S_{2g}\rangle$ vanishes because they have the same symmetry (g symmetry). The three singlet states are illustrated in Fig. 2.3 with transition moments and excitation energies. The SOS expression of γ_{iiii} in this model is thus reduced to

$$\gamma_{iiii} = \gamma_{iiii}^{\mathrm{II}} + \gamma_{iiii}^{\mathrm{III-2}} = -4 \frac{(\mu_{S_{1g},S_{1u}}^{i})^4}{(E_{S_{1g},S_{1u}})^3} + 4 \frac{(\mu_{S_{1g},S_{1u}}^{i})^2 (\mu_{S_{1u},S_{2g}}^{i})^2}{(E_{S_{1g},S_{1u}})^2 E_{S_{1g},S_{2g}}}. \quad (4.2.2)$$

Inserting Eqs. (2.3.1), (2.3.2), (2.3.7), (2.3.8), and (2.3.9) into (4.2.2), we obtain [9],

$$\begin{aligned} \gamma_{iiii} &= \gamma_{iiii}^{\mathrm{II}} + \gamma_{iiii}^{\mathrm{III-2}} \\ &= -\left(\frac{R_{\mathrm{BA}}^4}{U^3}\right) \frac{8(1-y)^4}{\left\{1 + \sqrt{1-(1-y)^2}\right\}^2 \left\{1 - 2r_K + \frac{1}{\sqrt{1-(1-y)^2}}\right\}^3} \\ &\quad + \left(\frac{R_{\mathrm{BA}}^4}{U^3}\right) \frac{4(1-y)^2}{\left\{1 - 2r_K + \frac{1}{\sqrt{1-(1-y)^2}}\right\}^2 \left\{\frac{1}{\sqrt{1-(1-y)^2}}\right\}}. \end{aligned} \quad (4.2.3)$$

The ND second hyperpolarizability $\gamma_{\mathrm{ND}\ iiii}$ is defined as

$$\begin{aligned} \gamma_{\mathrm{ND}\ iiii} &\equiv \frac{\gamma_{iiii}}{(R_{\mathrm{BA}}^4/U^3)} = \frac{\gamma_{iiii}^{\mathrm{II}}}{(R_{\mathrm{BA}}^4/U^3)} + \frac{\gamma_{iiii}^{\mathrm{III-2}}}{(R_{\mathrm{BA}}^4/U^3)} \\ &= -\frac{8(1-y)^4}{\left\{1 + \sqrt{1-(1-y)^2}\right\}^2 \left\{1 - 2r_K + \frac{1}{\sqrt{1-(1-y)^2}}\right\}^3} \\ &\quad + \frac{4(1-y)^2}{\left\{1 - 2r_K + \frac{1}{\sqrt{1-(1-y)^2}}\right\}^2 \left\{\frac{1}{\sqrt{1-(1-y)^2}}\right\}}. \end{aligned} \quad (4.2.4)$$

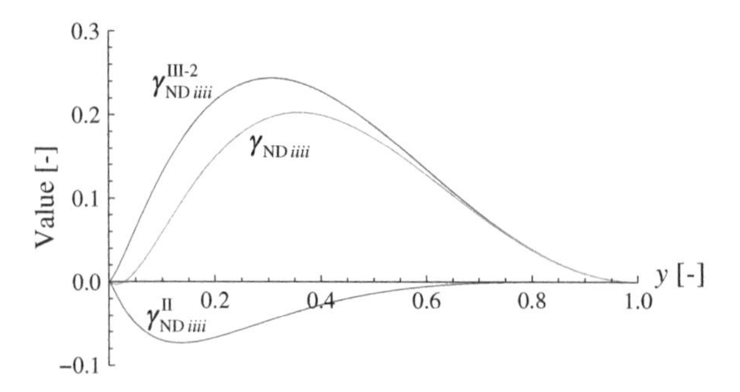

Fig. 4.2 Variations of nondimensional γ_{iiii} ($\gamma_{\mathrm{ND}iiii}$), $\gamma_{iiii}^{\mathrm{II}}$ ($\gamma_{\mathrm{ND}iiii}^{\mathrm{II}}$) and $\gamma_{iiii}^{\mathrm{III-2}}$ ($\gamma_{\mathrm{ND}iiii}^{\mathrm{III-2}}$) as a function of diradical character (y) in the case of $r_K = 0$

This expression indicates that $\gamma_{\mathrm{ND}\,iiii}$ is a function of the diradical character y and r_K. In the last equality of this equation, the first and second terms are the ND $\gamma_{iiii}^{\mathrm{II}}$ and $\gamma_{iiii}^{\mathrm{III-2}}$ ($\gamma_{\mathrm{ND}\,iiii}^{\mathrm{II}}$ and $\gamma_{\mathrm{ND}\,iiii}^{\mathrm{III-2}}$), respectively. Figure 4.2 shows the variations of $\gamma_{\mathrm{ND}\,iiii}^{\mathrm{II}}$, $\gamma_{\mathrm{ND}\,iiii}^{\mathrm{III-2}}$ and $\gamma_{\mathrm{ND}\,iiii}$ as the function of the diradical character y in the case of $r_K = 0$, which is approximately satisfied for most cases. The diradical character dependence of $\gamma_{\mathrm{ND}\,iiii}$ presents a bell-shape behavior with a maximum value at $y \sim 0.359$, which implies that $\gamma_{\mathrm{ND}\,iiii}$ is enhanced in the intermediate diradical character region [9]. This enhancement of $\gamma_{\mathrm{ND}\,iiii}$ originates from the contribution of type III-2 ($\gamma_{\mathrm{ND}\,iiii}^{\mathrm{III-2}}$), which is also enhanced in the intermediate diradical character region as shown in Fig. 4.2.

Next, the diradical character dependences of $\gamma_{\mathrm{ND}\,iiii}^{\mathrm{II}}$ and $\gamma_{\mathrm{ND}\,iiii}^{\mathrm{III-2}}$ are analyzed using the ND transition moments and excitation energies. From Eqs. (2.3.9) and (4.2.2), $\gamma_{\mathrm{ND}\,iiii}^{\mathrm{II}}$ is expressed by $\mu_{\mathrm{ND}\,S_{1g},S_{1u}}^{i}$ and $E_{\mathrm{ND}\,S_{1g},S_{1u}}$:

$$\gamma_{\mathrm{ND}\,iiii}^{\mathrm{II}} = -4\frac{\left(\mu_{\mathrm{ND}\,S_{1g},S_{1u}}^{i}\right)^4}{\left(E_{\mathrm{ND}\,S_{1g},S_{1u}}\right)^3}. \tag{4.2.5}$$

Figure 4.3a shows the diradical character dependences of the numerator $\left(\mu_{\mathrm{ND}\,S_{1g},S_{1u}}^{i}\right)^4$ and denominator $\left(E_{\mathrm{ND}\,S_{1g},S_{1u}}\right)^3$ of Eq. (4.2.5), and $\gamma_{\mathrm{ND}\,iiii}^{\mathrm{II}}$ in the case of $r_K = 0$. In the case of $y \to 0$, the numerator approaches a finite value (0.25), while the denominator does infinity, which results in $\gamma_{\mathrm{ND}\,iiii}^{\mathrm{II}} = 0$ at $y = 0$. On the other hand, in the case of $y \to 1$, the denominator and numerator approach a finite value and 0, respectively, leading to $\gamma_{\mathrm{ND}\,iiii}^{\mathrm{II}} \to 0$. Although both of the denominator $\left(E_{\mathrm{ND}\,S_{1g},S_{1u}}\right)^3$ and numerator $\left(\mu_{\mathrm{ND}\,S_{1g},S_{1u}}^{i}\right)^4$ decrease as increasing y from 0 to 1, $\left(E_{\mathrm{ND}\,S_{1g},S_{1u}}\right)^3$ presents a rapid decrease in the small y region as compared to $\left(\mu_{\mathrm{ND}\,S_{1g},S_{1u}}^{i}\right)^4$, which causes the enhancement of $\gamma_{\mathrm{ND}\,iiii}^{\mathrm{II}}$ in the small diradical character region. $\gamma_{\mathrm{ND}\,iiii}^{\mathrm{II}}$ takes a minimum (~ -0.074) at $y \approx 0.133$.

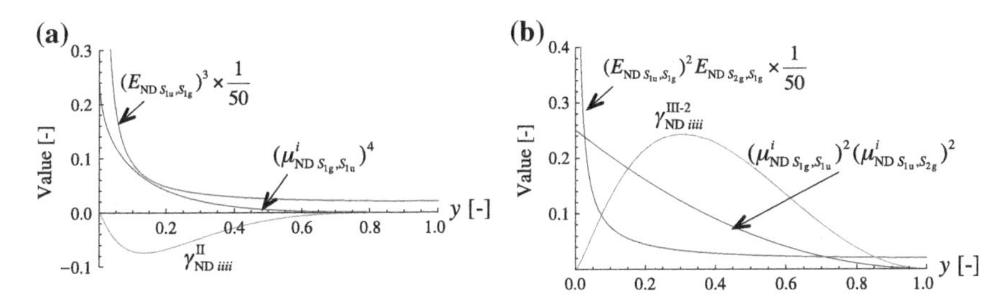

Fig. 4.3 Diradical character (y) dependences of $\gamma^{\mathrm{II}}_{\mathrm{ND}iiii}$, $(\mu^i_{\mathrm{ND}S_{1g},S_{1u}})^4$ and $(E_{\mathrm{ND}S_{1u},S_{1g}})^3$ (**a**), and those of $\gamma^{\mathrm{III-2}}_{\mathrm{ND}iiii}$, $(\mu^i_{\mathrm{ND}S_{1g},S_{1u}})^2(\mu^i_{\mathrm{ND}S_{1u},S_{2g}})^2$ and $(E_{\mathrm{ND}S_{1u},S_{1g}})^2 E_{\mathrm{ND}S_{2g},S_{1g}}$ (**b**) in the case of $r_K = 0$

From Eqs. (2.3.9) and (4.2.2), $\gamma^{\mathrm{III-2}}_{\mathrm{ND}\,iiii}$ is expressed by $\mu^i_{\mathrm{ND}\,S_{1g},S_{1u}}$, $\mu^i_{\mathrm{ND}\,S_{1u},S_{2g}}$, $E_{\mathrm{ND}\,S_{1u},S_{1g}}$ and $E_{\mathrm{ND}\,S_{2g},S_{1g}}$ as

$$\gamma^{\mathrm{III-2}}_{\mathrm{ND}\,iiii} = 4\frac{\left(\mu^i_{\mathrm{ND}\,S_{1g},S_{1u}}\right)^2\left(\mu^i_{\mathrm{ND}\,S_{1u},S_{2g}}\right)^2}{\left(E_{\mathrm{ND}\,S_{1u},S_{1g}}\right)^2 E_{\mathrm{ND}\,S_{2g},S_{1g}}}. \tag{4.2.6}$$

The variations in the numerator $\left(\mu^i_{\mathrm{ND}\,S_{1g},S_{1u}}\right)^2\left(\mu^i_{\mathrm{ND}\,S_{1u},S_{2g}}\right)^2$ and denominator $\left(E_{\mathrm{ND}\,S_{1u},S_{1g}}\right)^2 E_{\mathrm{ND}\,S_{2g},S_{1g}}$ of Eq. (4.2.6), and $\gamma^{\mathrm{III-2}}_{\mathrm{ND}\,iiii}$ in the case of $r_K = 0$ are shown as a function of y in Fig. 4.3b. The features of Fig. 4.3b at the limits of $y = 0$ and 1 are very similar to those of Fig. 4.3a: the denominator and numerator approach infinity and a finite value, respectively, as $y \to 0$, resulting in $\gamma^{\mathrm{III-2}}_{\mathrm{ND}\,iiii} \to 0$, while they do a finite value and 0, respectively, as $y \to 1$, leading to $\gamma^{\mathrm{III-2}}_{\mathrm{ND}\,iiii} \to 0$ again. Both of the denominator and numerator decrease as increasing the diradical character from 0 to 1. However, the denominator decreases rapidly in the small diradical character region, while the numerator shows a slow decrease, which results in the enhancement of $\gamma^{\mathrm{III-2}}_{\mathrm{ND}\,iiii}$ in the intermediate diradical character region: $\gamma^{\mathrm{III-2}}_{\mathrm{ND}\,iiii}$ takes a maximum ($\sim$0.243) at $y \sim 0.306$.

The reason for $\left|\gamma^{\mathrm{III-2}}_{\mathrm{ND}\,iiii}\right| > \left|\gamma^{\mathrm{II}}_{\mathrm{ND}\,iiii}\right|$ except for $y \sim 0$ is that the numerator of Eq. (4.2.5), $\left(\mu^i_{\mathrm{ND}\,S_{1g},S_{1u}}\right)^4$, decreases more rapidly than that of Eq. (4.2.6) $\left(\mu^i_{\mathrm{ND}\,S_{1g},S_{1u}}\right)^2\left(\mu^i_{\mathrm{ND}\,S_{1u},S_{2g}}\right)^2$ (see Fig. 4.3). As shown in Fig. 2.3, $(\mu^i_{S_{1g},S_{1u}})^2$ and $(\mu^i_{S_{1u},S_{2g}})^2$ present decrease and increase, respectively, with the increase in the diradical character (these features can be understood from the neutral and ionic components of the ground and excitation states as mentioned in Sect. 2.3). Therefore, $\left(\mu^i_{\mathrm{ND}\,S_{1g},S_{1u}}\right)^2\left(\mu^i_{\mathrm{ND}\,S_{1u},S_{2g}}\right)^2$ keeps a larger value than $\left(\mu^i_{\mathrm{ND}\,S_{1g},S_{1u}}\right)^4$ in the whole diradical character region.

Finally, another type of description of $\gamma_{\mathrm{ND}\,iiii}$ and the diradical character is provided in relation to the effective exchange integral J [9]. This is useful for clarifying the correlation between NLO property and magnetic interaction for open-shell molecules with various diradical characters. In the case of two-electron systems in the Heisenberg model [10], the energy gap between the singlet and triplet states is in agreement with $2J$. From Eqs. (2.1.9) and (2.1.11b), we obtain

$$2J \equiv {}^{1}E_{1\mathrm{g}} - {}^{3}E_{1\mathrm{u}} = 2K_{ab} + \frac{U - \sqrt{U^2 + 16t_{ab}^2}}{2}. \tag{4.2.7}$$

If we define the nondimensional effective exchange integral as $r_J \equiv 2J/U$,

$$r_J = r_K + \frac{1 - \sqrt{1 + 16r_t^2}}{2}. \tag{4.2.8}$$

Using this equation, r_t can be given as a function of r_K and r_J,

$$r_t = \frac{\sqrt{(r_K - r_J)(r_K - r_J + 1)}}{2}. \tag{4.2.9}$$

Inserting Eq. (4.2.9) into Eq. (2.2.11), we obtain the diradical character as a function of r_K and r_J [9],

$$y = 1 - \frac{2\sqrt{(r_K - r_J)(r_K - r_J + 1)}}{1 + 2(r_K - r_J)}. \tag{4.2.10}$$

Furthermore, inserting Eq. (4.2.9) into Eqs. (2.3.1), (2.3.2), (2.3.7), and (2.3.8) gives the following formula [9]:

$$\left(\mu^i_{S_{1\mathrm{g}},S_{1\mathrm{u}}}\right)^2 = R_{\mathrm{BA}}^2 \frac{r_K - r_J}{1 + 2(r_K - r_J)}, \tag{4.2.11}$$

$$\left(\mu^i_{S_{1\mathrm{u}},S_{2\mathrm{g}}}\right)^2 = \frac{R_{\mathrm{BA}}^2}{2}\left\{1 + \frac{1}{1 + 2(r_K - r_J)}\right\}, \tag{4.2.12}$$

$$E_{S_{1\mathrm{u}},S_{1\mathrm{g}}} = U(1 - r_J), \tag{4.2.13}$$

and

$$E_{S_{2\mathrm{g}},S_{1\mathrm{g}}} = U\{1 + 2(r_K - r_J)\}, \tag{4.2.14}$$

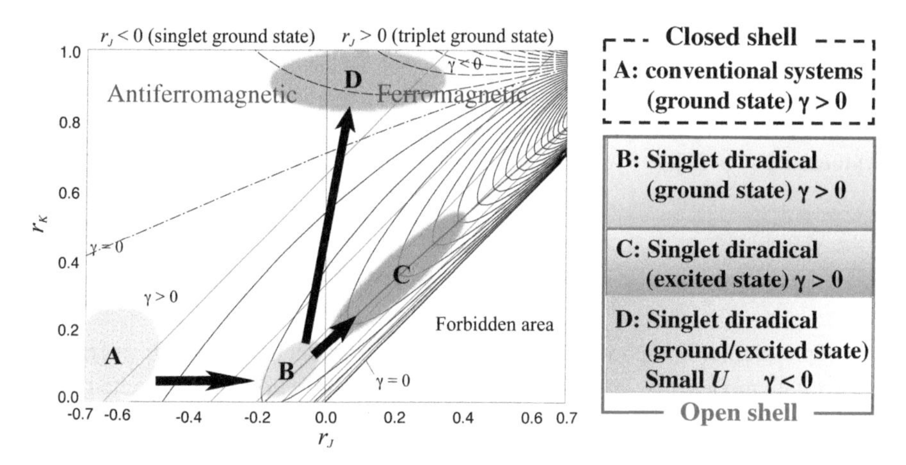

Fig. 4.4 Contours of $\gamma_{\text{ND}iiii}$ ($-2.0 \leq \gamma_{\text{ND}iiii} < 2.0$, interval $= 0.1$) on the plane (r_J, r_K). Positive and negative r_J regions indicate the systems with antiferromagnetic and ferromagnetic ground states, respectively. *Solid*, *dotted*, and *dashedblack lines* represent positive, negative, and zero lines of $\gamma_{\text{ND}iiii}$, respectively. *Solid green* and *red lines* represent the iso-*y*-lines and the *ridge line* connecting the (r_J, r_K) points exhibiting maximum $\gamma_{\text{ND}iiii}$ values, respectively. Region **A** (closed-shell) and **B–D** (open-shell) are also shown. Region **A** represents conventional NLO systems, while **B–D** are theoretically predicted, where the *arrow* indicates the order of exploration

From Eqs. (4.2.11)–(4.2.14) and (4.2.2), we obtain $\gamma_{\text{ND}iiii}$ as a function of r_K and r_J [9],

$$\begin{aligned}\gamma_{\text{ND}iiii} &= \frac{\gamma}{\left(R_{\text{BA}}^2/U^3\right)} \\ &= -\frac{4(r_K - r_J)^2}{\{1 + 2(r_K - r_J)\}^2(1 - r_J)^3} + \frac{4(r_K - r_J + 1)(r_K - r_J)}{\{1 + 2(r_K - r_J)\}^3(1 - r_J)^2}. \end{aligned} \tag{4.2.15}$$

Figure 4.4 shows the variation of $\gamma_{\text{ND}iiii}$ (black lines) in the (r_J, r_K) plane along with the iso-y-line (green lines). It is found that the ridge line, which connects the (r_J, r_K) points exhibiting maximum $\gamma_{\text{ND}iiii}$ values, is almost parallel to the iso-y-lines between $y = 0.3$ and 0.5. This feature supports the enhancement of $\gamma_{\text{ND}iiii}$ in the intermediate y region in the case with finite r_K values, similar to the case with $r_K = 0$. The systems having open-shell singlet ground states (region **B**) are found to exhibit larger $\gamma_{\text{ND}iiii}$ values than the conventional closed-shell systems (region **A**). Furthermore, the maximum $\gamma_{\text{ND}iiii}$ increases from the lower left to the upper right corner on the (r_J, r_K) plane, i.e., by increasing r_K and r_J. In particular, region **C**, which indicates the systems with triplet ground states and singlet excited states having intermediate diradical characters, presents further enhancement of $\gamma_{\text{ND}iiii}$ in the singlet excited states as compared to region **B**. Region **C** corresponds to the ferromagnetic interaction region ($J > 0$), which has been actively investigated in an effort to realize molecular magnets. There is another interesting region **D**, which presents negative $\gamma_{\mathrm{ND}iiii}$ values with large amplitudes and has singlet or triplet ground states. Although regions **C** and **D** are attractive and should be investigated in the future, we focus on the compounds belonging to region **B** because several real compounds belonging to region **B** have been discovered or synthesized [11–26].

4.3 One-Photon Absorption for Symmetric Systems

The expression of one-photon absorption (OPA) cross section at photon energy $\hbar\omega$ of two state models ($|S_{1g}\rangle$: ground state, $|S_{1u}\rangle$: optically-allowed first excited state) for symmetric diradical systems is given by [27]

$$\sigma^{(1)}(\omega) = \left(\frac{4\pi}{3\hbar cn}\right)(\hbar\omega)\mu^2_{S_{1u},S_{1g}}\left(\frac{\hbar\Gamma_{S_{1u},S_{1g}}}{(E_{S_{1u},S_{1g}} - \hbar\omega)^2 + (\hbar\Gamma_{S_{1u},S_{1g}})^2}\right), \tag{4.3.1}$$

where n is the refractive index of the sample, and $\Gamma_{S_{1u},S_{1g}}$, the damping factor from state S_{1u} to S_{1g}, is assumed to satisfy the relation [27]:

$$\hbar\Gamma_{S_{1u},S_{1g}} = f_{S_{1u},S_{1g}}E_{S_{1u},S_{1g}}. \tag{4.3.2}$$

The $f_{S_{1u},S_{1g}}$ $(0 \le f_{S_{1u},S_{1g}} \le 1)$ coefficient is defined to describe the excitation energy dependence of the damping factor, which indicates that the lower the excited state, the longer the lifetime of the state. In particular, using Eqs. (2.3.1), (2.3.7), (4.3.1) and (4.3.2), the nondimensional OPA peak cross section $\sigma^{(1)}_{\mathrm{ND}}(\omega_{S_{1u},S_{1g}})$, where $\hbar\omega_{S_{1u},S_{1g}} = E_{S_{1u},S_{1g}}$, is expressed by [27]

$$\sigma^{(1)}_{\mathrm{ND}}(\omega_{S_{1u},S_{1g}}) \equiv \frac{\sigma^{(1)}(\omega_{S_{1u},S_{1g}})}{\left(\frac{4\pi}{3\hbar cn}\right)(\mathrm{e}R_{\mathrm{BA}})^2} = \frac{1 - \sqrt{1-(1-y)^2}}{2f_{S_{1u},S_{1g}}}, \tag{4.3.3}$$

where the peak position is represented by Eqs. (2.3.1) and (2.3.9):

$$E_{\mathrm{ND}\;S_{1u},S_{1g}} \equiv \frac{E_{S_{1u},S_{1g}}}{U} = \frac{1}{2}\left\{1 - 2r_K + \frac{1}{\sqrt{1-(1-y)^2}}\right\}. \tag{4.3.4}$$

This equation indicates that the peak intensity $\sigma^{(1)}_{\mathrm{ND}}(\omega_{S_{1u},S_{1g}})$ decreases in inverse proportion to the damping coefficient, $f_{S_{1u},S_{1g}}$, and also decreases with increasing diradical character y if keeping $f_{S_{1u},S_{1g}}$ constant as shown in Fig. 4.5a. Also, in case of comparison with systems of different size R_{BA}, one should pay attention to the fact that $\sigma^{(1)}(\omega_{S_{1u},S_{1g}})$ [not nondimensional $\sigma^{(1)}_{\mathrm{ND}}(\omega_{S_{1u},S_{1g}})$] is proportional to R^2_{BA}.

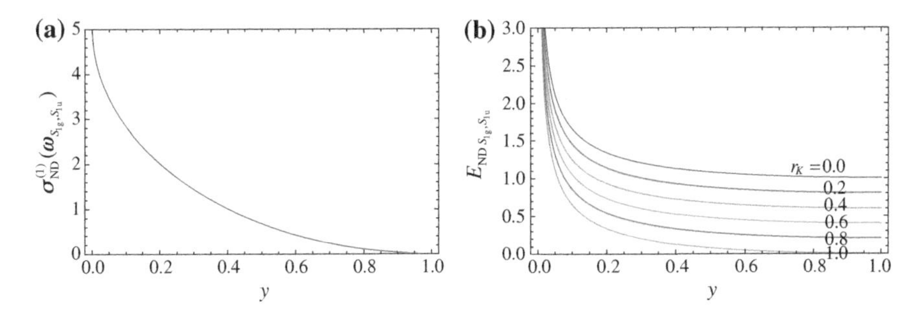

Fig. 4.5 Diradical character dependence of OPA peak intensity $\sigma_{\mathrm{ND}}^{(1)}(\omega_{S_{1u},S_{1g}})$ (**a**) and y-dependences of excitation energy $E_{\mathrm{ND}\,S_{1u},S_{1g}}$ with different r_K (**b**)

As seen from Eq. (4.3.4), the OPA peak position moves to the lower energy region as increasing the diradical character and thus converges to a value, $1 - r_K$, which shows the decrease of the nondimensional excitation energy with the increase of r_K as shown in Fig. 4.5b. This behavior has to be contrasted with the well-known feature that closed-shell π-conjugated hydrocarbon systems exhibits an increase of OPA intensity and a decrease of excitation energy as increasing the π-conjugation. It is also noted that the real peak position $E_{S_{1u},S_{1g}}$ has the possibility of increasing again in the large y region, where $E_{\mathrm{ND}\,S_{1u},S_{1g}}$ converges to a stationary value, because $E_{S_{1u},S_{1g}}$ is also proportional to U, and y tends to increase with the increase in U [see Eq. (2.2.10)]. Such unique feature is predicted in several open-shell singlet molecules with large y values [28]. Indeed, this relationship in the OPA peak intensity and position as a function of the diradical character can be used for qualitatively estimating the diradical character [29].

4.4 Two-Photon Absorption for Symmetric Systems

The two-photon absorption (TPA) cross section of the two-site diradical model system at photon energy $\hbar\omega_{\mathrm{ph}}$ of the applied laser field is given by [30, 31],

$$\sigma^{(2)}(\omega_{\mathrm{ph}}) = \left(\frac{16\pi^2}{5c^2n^2\hbar}\right)(\hbar\omega_{\mathrm{ph}})^2 L^4 \left\{\frac{\mu_{S_{1g},S_{1u}}^2 \mu_{S_{1u},S_{2g}}^2}{\left(E_{S_{1u},S_{1g}} - \hbar\omega_{\mathrm{ph}}\right)^2 + \Gamma_{S_{1u},S_{1g}}^2)}\right\} \times \left\{\frac{\Gamma_{S_{2g},S_{1g}}}{\left(E_{S_{2g},S_{1g}} - 2\hbar\omega_{\mathrm{ph}}\right)^2 + \Gamma_{S_{2g},S_{1g}}^2}\right\}, \tag{4.4.1}$$

where n is the refractive index of the sample and L is the local field factor. The damping factor from state β to α, $\Gamma_{\beta\alpha}$, is assumed to satisfy the relation [32]:

$$\Gamma_{\beta\alpha} = f_{\beta\alpha} E_{\beta\alpha}, \tag{4.4.2}$$

where the coefficient $f_{\beta\alpha}(0 \le f_{\beta\alpha} \le 1)$ is defined to describe the excitation energy dependence of the damping factor:

$$\frac{\Gamma_{S_{1u},S_{1g}}}{\Gamma_{S_{2g},S_{1g}}} = \frac{f_{S_{1u},S_{1g}} E_{S_{1u},S_{1g}}}{f_{S_{2g},S_{1g}} E_{S_{2g},S_{1g}}} \equiv p r_E. \tag{4.4.3}$$

where r_E indicates the excitation energy ratio $E_{S_{1u},S_{1g}}/E_{S_{2g},S_{1g}}$ [the $0 \le r_E \le 1$ relationship is considered in this study; see Eqs. (2.3.1) and (2.3.2) as well as $r_K \ge r_J$ [Eq. (4.2.8)] and $r_K \ge 0$] and p is the coefficient ratio $f_{S_{1u},S_{1g}}/f_{S_{2g},S_{1g}}$ $(0 \le p \le 1)$. This relation implies that the lower the excited state, the longer the lifetime of the state, i.e., the lifetime of the lowest excited state ($|S_{1u}\rangle$) is assumed to be larger than that of the higher excited state ($|S_{2g}\rangle$). Using Eqs. (2.3.1), (2.3.2), (2.3.7), (2.3.8), (4.2.10) and Eqs. (4.4.1)–(4.4.3), the nondimensional TPA peak cross section satisfying $2\hbar\omega_{\mathrm{ph}} = E_{S_{2g},S_{1g}}$ is represented by [32]

$$\sigma_{\mathrm{ND}}^{(2)}(\omega_{S_{2g},S_{1g}}/2) \equiv \frac{\sigma^{(2)}(\omega_{S_{2g},S_{1g}}/2)}{\left(\frac{16\pi^2}{5c^2n^2\hbar}\right) L^4 \frac{(eR_{\mathrm{BA}})^4}{U}}$$

$$= \frac{q^2}{4f_{S_{2g},S_{1g}}\sqrt{1-q^2}\left\{(1-2r_K)^2 + p^2 f_{S_{2g},S_{1g}}^2 \left(1 - 2r_K + \frac{1}{\sqrt{1-q^2}}\right)^2\right\}} \tag{4.4.4a}$$

$$= \frac{q^2}{4f_{S_{2g},S_{1g}}\sqrt{1-q^2}\left\{\left[2(1-r_J) - \frac{1}{\sqrt{1-q^2}}\right]^2 + 4p^2 f_{S_{2g},S_{1g}}^2 (1-r_J)^2\right\}} \tag{4.4.4b}$$

$$= \frac{(r_K - r_J)(r_K - r_J + 1)}{f_{S_{2g},S_{1g}}\left\{4p^2 f_{S_{2g},S_{1g}}^2 (r_J - 1)^2 + (1-2r_K)^2\right\}\{2(r_K - r_J) + 1\}}, \tag{4.4.4c}$$

where $q = 1 - y$ (effective bond order) and Eqs. (4.4.4a), (4.4.4b) and (4.4.4c) show the dependence of $\sigma_{\mathrm{ND}}^{(2)}(\omega_{S_{2g},S_{1g}}/2)$ on $(q, r_K, p, f_{S_{2g},S_{1g}})$, $(q, r_J, p, f_{S_{2g},S_{1g}})$ and $(r_J, r_K, p, f_{S_{2g},S_{1g}})$, respectively. Here, Eq. (4.4.4a) is employed for clarifying the diradical character dependence of $\sigma_{\mathrm{ND}}^{(2)}(\omega_{S_{2g},S_{1g}}/2)$. We firstly consider the dependences of $\sigma_{\mathrm{ND}}^{(2)}(\omega_{S_{2g},S_{1g}}/2)$ on y and r_K by setting $(p, f_{S_{2g},S_{1g}}) = (1, 0.1)$, which qualitatively reproduces the amplitudes of experimental damping factors [30, 31, 33, 34]. Figure 4.6a shows the behavior of $\sigma_{\mathrm{ND}}^{(2)}(\omega_{S_{2g},S_{1g}}/2)$ on the y–r_K plane [Eq. (4.4.4a)] together with the isolines of excitation energy ratio r_E [Eqs. (2.3.1)

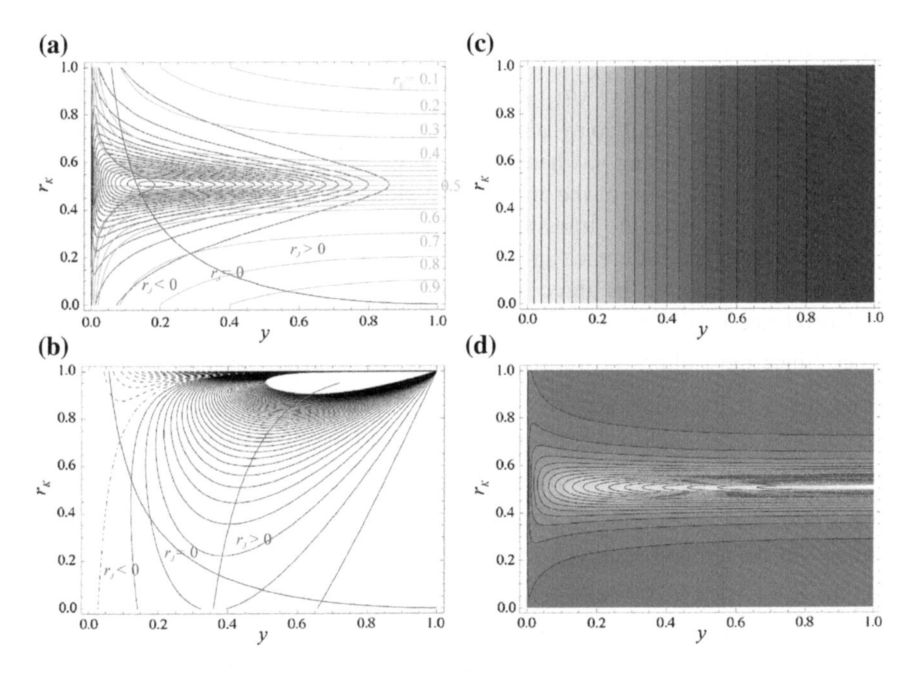

Fig. 4.6 Contours (*solid black lines*) of $\sigma^{(2)}_{\mathrm{ND}}(\omega_{S_{2g},S_{1g}}/2)$ using $(p, f_{S_{2g},S_{1g}}) = (1,\ 0.1)$ in Eq. (4.4.4a) ($0.0 \leq \sigma^{(2)}_{\mathrm{ND}}(\omega_{S_{2g},S_{1g}}/2) \leq 100.0$, interval = 5.0) (**a**), of γ_{ND} (**b**), of transition moment term $(\mu_{S_{1g},S_{1u}}/eR_{\mathrm{BA}})^2(\mu_{S_{1u},S_{2g}}/eR_{\mathrm{BA}})^2$ (0–0.25, interval 0.01) (**c**), and of energy term $\sigma^{(2)}_{\mathrm{ND}}(\omega_{S_{2g},S_{1g}}/2)/[(\mu_{S_{1g},S_{1u}}/eR_{\mathrm{BA}})^2(\mu_{S_{1u},S_{2g}}/eR_{\mathrm{BA}})^2]$ (0–1,000, interval 50) (**d**). In **a**, *solid green lines* indicate iso-r_E lines. In **a** and **b**, a *blue line* indicates $r_J = 0$ line, and a *red line* represents the *ridge line* connecting the (y, r_K) points exhibiting maximum values [$\sigma^{(2)}_{\mathrm{ND}}(\omega_{S_{2g},S_{1g}}/2)$ for **a** and γ_{ND} for **d**] at each r_K. In **b**, *dot-dashed* and *dashed black lines* represent zero and negative lines of γ_{ND}, respectively. In (**c**) and (**d**), lighter regions represent larger values

and (2.3.2)], indicating the degree of double resonance between the one- and two-photon transitions, which causes the one-photon resonance enhancement of the TPA peak near $r_E = 0.5$ (achieved by $q = 1 - y = 0.5$) [30–32] in addition to the sequential TPA process. The $r_J = 0$ line [Eqs. (4.2.8) and (4.2.9)] enables us to clarify the ground-state magnetic interactions of the system: the upper and lower regions with respect to $r_J = 0$ line correspond to the ferro-($r_J > 0$) and antiferro-($r_J < 0$) magnetic regions, respectively. For any r_K, the maximum in $\sigma^{(2)}_{\mathrm{ND}}(\omega_{S_{2g},S_{1g}}/2)$ appears in small but finite y regions. Moreover, this maximum as well as the corresponding y value also increase when approaching $r_K = 0.5$ (see the ridge curve connecting the points with maximum $\sigma^{(2)}_{\mathrm{ND}}(\omega_{S_{2g},S_{1g}}/2)$ in Fig. 4.6a). The iso-$\sigma^{(2)}_{\mathrm{ND}}(\omega_{S_{2g},S_{1g}}/2)$ lines appear to reflect the iso-r_E lines, though the $\sigma^{(2)}_{\mathrm{ND}}(\omega_{S_{2g},S_{1g}}/2)$ values approach zero values for $y \to 0$ and 1. This can be understood by partitioning $\sigma^{(2)}_{\mathrm{ND}}(\omega_{S_{2g},S_{1g}}/2)$ into the transition moment and the energy terms defined, respectively, by $(\mu_{S_{1g},S_{1u}}/eR_{BA})^2(\mu_{S_{1u},S_{2g}}/eR_{BA})^2$ and $\sigma_{ND}^{(2)}(E_{S_{2g},S_{1g}}/2)/\{(\mu_{S_{1g},S_{1u}}/eR_{BA})^2(\mu_{S_{1u},S_{2g}}/eR_{BA})^2\}$ (see Fig. 4.6c and d). The transition moment term depends only on y, and nonlinearly increases from 0 (at $y = 1$) to 1/4 (at $y = 0$) with decreasing y, whereas the energy term provides similar isolines to r_E except for the increase of the energy term along the $r_K = 0.5$ line as a function of y. The variations in the transition moment and energy terms with respect to y originate from the difference in the relative contributions of the diradical and ionic configurations in the three singlet states, S_{1g}, S_{1u} and S_{2g} [9], while the r_K-dependence of the energy term is mostly described by the double resonance effect. Since the isolines of $\sigma_{ND}^{(2)}(\omega_{S_{2g},S_{1g}}/2)$ result from the product of these two terms, the (y, r_K)-dependences of $\sigma_{ND}^{(2)}(\omega_{S_{2g},S_{1g}}/2)$ in small and large y regions are governed by the energy and the transition moment terms, respectively.

As to the relationship between $\sigma_{ND}^{(2)}(\omega_{S_{2g},S_{1g}}/2)$ and the ground-state magnetic interaction, it is noteworthy that the maximum $\sigma_{ND}^{(2)}(\omega_{S_{2g},S_{1g}}/2)$ value ($\sim$96.9) is realized in the ground-state ferromagnetic region $(y, r_K) \sim (0.184, 0.504)$, and that the $\sigma_{ND}^{(2)}(\omega_{S_{2g},S_{1g}}/2)$ values in the intermediate y region are maximized around $r_K = 0.5$ ($r_E = 0.5$) in the ground-state ferromagnetic region ($r_J > 0$), and their values are larger than those in the region near $y = 0$ in the ground-state antiferromagnetic region ($r_J < 0$). Namely, the TPA resonance enhancement is larger for singlet diradical systems with relatively small diradical characters than for closed-shell ($y = 0$) and pure diradical ($y \sim 1$) systems, but it is even larger for excited-state singlets with intermediate diradical characters in the ferromagnetic interaction region ($r_J > 0$). This feature is thus in qualitative conformity with the variations of the nondimensional static γ (γ_{ND}) [9] (see also Fig. 4.6b). Indeed, the ridge line connecting the (y, r_K) points leading to the maximum γ_{ND} in the intermediate y region shows that larger γ values are obtained when increasing simultaneously y and r_K. Nevertheless, the maximum γ_{ND} values occur for larger diradical character than for $\sigma_{ND}^{(2)}(\omega_{S_{2g},S_{1g}}/2)$. This difference originates from the $(\hbar\omega_{ph})^2[= (E_{S_{2g},S_{1g}}/2)^2 \propto 1/(1 - q^2)]$ factor in Eq. (4.4.1). Such variations of γ_{ND} and $\sigma_{ND}^{(2)}(\omega_{S_{2g},S_{1g}}/2)$ in open-shell molecules as a function of the diradical character have to be distinguished from their monotonic increases observed in closed-shellquadrupolar systems when increasing the donor and acceptor strengths [35].

By taking advantage of the VCI parameters of *s*-indaceno[1,2,3-*cd*;6,7,8-*c′d′*]diphenalene (IDPL) and *as*-indaceno[1,2,3-*cd*;6,7,8-*c′d′*]diphenalene (*as*-IDPL) determined from UNOCASCI(2,2,)/6-31G* calculations [9] and by setting $(p, f_{S_{2g},S_{1g}})$ to typical values of (1, 0.1), the TPA cross section ratio attains $\sim$13 in favor of IDPL, which displays an intermediate diradical character ($y = 0.54$) while *as*-IDPL is better described as a pure diradical ($y \sim 0.83$) (see Table 4.1). Similarly, the ratio of TPA cross sections is large ($\sim$14) when comparing IDPL to pentacene, in good agreement with the enhancement of TPA cross sections reported in Ref. [30]. In the latter case, the UNOCASCI(2,2)/6-31G* calculation,

Table 4.1 VCI parameters and TPA cross sections, $\sigma_{\mathrm{ND}}^{(2)}(E_{S_{2g},S_{1g}}/2)U/(eR_{\mathrm{BA}})^4$ [Eq. (4.4.4)], using UNOCASCI(2,2)/6-31G* calculation results and $(p, f_{S_{2g},S_{1g}}) = (1, 0.1)$ for indaceno[1,2,3-*cd*;6,7,8-*c*′*d*′]diphenalene (IDPL), *as*-indaceno[1,2,3-*cd*;6,7,8-*c*′*d*′]diphenalene (*as*-IDPL) and pentacene

	IDPL	*as*-IDPL	pentacene
r_J	−0.061	−0.001	−3.880
r_K	0.004	0.007	0.003
r_t	0.131	0.043	2.189
y	0.537	0.830	0.006
$\sigma_{\mathrm{ND}}^{(2)}(E_{S_{2g},S_{1g}}/2)U/(eR_{\mathrm{BA}})^4 \times [10^2$ a.u.]	2,895	229	212

carried out using the HOMO − 2 and LUMO + 2 to describe the dominant tensor component, leads indeed to the very small diradical character of 0.006.

4.5 First Hyperpolarizability (β) for Asymmetric Systems

Using the results of asymmetric diradical systems in Chap. 3, we here examine the static β along the bond axis, which is a good approximation to the off-resonant β. The perturbative expression of β includes virtual excitation processes referred to as types I(b1) and II(b2), which disappear for symmetric molecular systems but not necessarily for asymmetric ones [36]. In the case of an asymmetric two-site diradical model with three singlet states {g, k, f}, the expression reduces to $\beta = \beta^{\mathrm{I}} + \beta^{\mathrm{II}}$, where type I and II contributions are given by [1–5]

$$\beta^{\mathrm{I}} = 3\frac{(\mu_{\mathrm{gk}})^2 \Delta\mu_{\mathrm{kk}}}{(E_{\mathrm{kg}})^2} + 3\frac{(\mu_{\mathrm{gf}})^2 \Delta\mu_{\mathrm{ff}}}{(E_{\mathrm{fg}})^2} \text{ and } \beta^{\mathrm{II}} = 6\frac{\mu_{\mathrm{gk}}\mu_{\mathrm{kf}}\mu_{\mathrm{fg}}}{E_{\mathrm{kg}}E_{\mathrm{fg}}}. \tag{4.5.1}$$

The first and second terms in β^{I} are referred to as b11(g-k-k-g) and b12(g-f-f-g), respectively, while type II contribution as b2(g-k-f-g). In the following, a nondimensional β, defined by $\beta/(R^3/U^2)$, is investigated. Figure 4.7a and b show the evolution of these different contributions as a function of y_{S} for $r_h = 0.4$ and 0.8. In both cases, the dominant contribution is positive (in the same direction as the ground state dipole moment) and comes from b11(g-k-k-g), whereas when $y_{\mathrm{S}} < 0.5$, b2(g-k-f-g) is negative and slightly impacts the total value. In addition, it is found that β shows a bell-shape variation as a function of y_{S}, and that the maximum β (β_{max}) amplitude is significantly enhanced when increasing r_h, and the corresponding y_{S}, referred to as y_{Smax}, moves to higher y_{S} region (see Fig. 4.8a). These behaviors of β as a function of y_{S} are understood by the dependences of the related excitation energies and excitation properties as a function of y_{S} and r_h

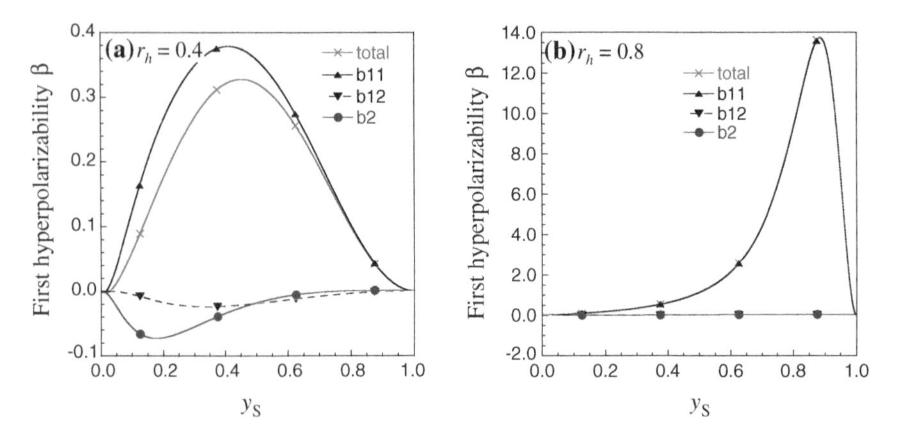

Fig. 4.7 y_S dependences of nondimensional β [total, b11(g-k-k-g), b12(g-f-f-g), and b2(g-k-f-g)] for $r_h = 0.4$ (**a**) and 0.8 (**b**)

shown in Fig. 3.3 together with Eq. (4.5.1). Indeed, b11(g-k-k-g) is dominant because, for any y_S and r_h, $|\mu_{gk}|/R > |\mu_{gf}|/R$ and $E_{kg} < E_{fg}$, while the negative $\Delta\mu_{ff}/R$ with relatively large amplitudes gives the slight negative contribution of b12 in the small y_S region.

It is found from Fig. 3.3d–f that E_{kg} decreases around $r_h = 1$, E_{fg} increases monotonically with r_h, $|\mu_{gk}|/R$ is maximized around $r_h = 1$, and $\Delta\mu_{kk}/R$, which has a large amplitude except for $r_h = 0$ and 1, changes sign from positive to negative at $r_h \sim 1$. The increase in y_S is shown to amplify these features (see Fig. 3.3d–f). This leads to the features: (i) the sign change of β at $r_h \sim 1$ (in the present case, positive for $r_h < 1$, while negative for $r_h > 1$), (ii) the increase/decrease of $|\beta_{max}|$ with r_h for $(r_h < 1)/(r_h > 1)$, and (iii) the associated effect on y_{Smax}, which increases/decreases with r_h for $(r_h < 1)/(r_h > 1)$. These features are exemplified by the y_S–β plots for different r_h as shown in Fig. 4.8a and b, e.g., (y_{Smax}, β_{max}) = (0. 327, 0.453) for $r_h = 0.4$, (0.676, 1.48) for $r_h = 0.7$, (0.886, 13.7) for $r_h = 0.9$, (0. 885, −14.2) for $r_h = 1.1$, (0.666, −1.63) for $r_h = 1.3$, and (0.410, −0.431) for $r_h = 1.6$. Then, from the y_A–β plots (Fig. 4.8c) for $r_h < 1$, β attains a maximum in

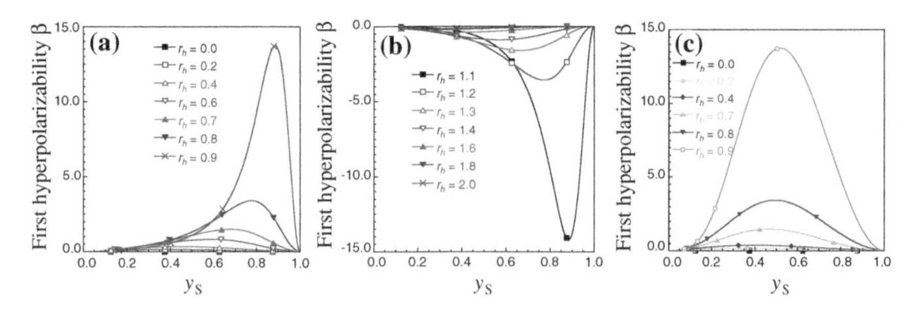

Fig. 4.8 Diradical character dependences of nondimensional β as a function of r_h: $y_S - \beta$ for r_h 0.0 − 0.9 (**a**) and 1.1 − 2.0 (**b**) as well as $y_A - \beta$ for $r_h = 0.0 - 0.9$ (**c**)

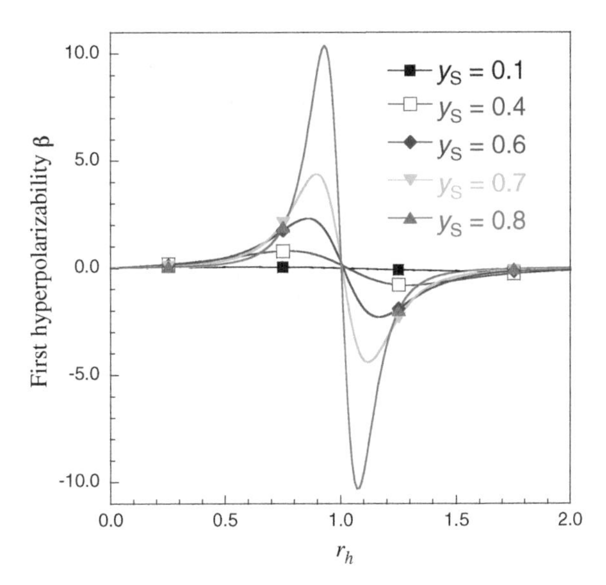

Fig. 4.9 r_h dependences of nondimensional β for y_S $(0.1 \leq y_S \leq 0.8)$

the intermediate y_A region for any r_h, and β_{max} increases with r_h. The asymmetricity (r_h) dependence of β, shown in Fig. 4.9 for different y_S, is qualitatively the same for all y_S values: when r_h increases, β increases, attains a maximum, decreases, changes its sign, attains a minimum, and then increases until reaching zero. This r_h-behavior is similar to that found for β in closed-shell prototypical merocyanine chromophores under a static electric field [37, 38], which similarly causes asymmetric electron distributions. In addition, this study shows that upon increasing y_S, the β amplitudes are enhanced and the β peak position for each y_S becomes closer to $r_h = 1$, e.g., $\beta_+(\beta_-) = 0.787(-0.814)$ at $r_h = 0.770(1.29)$ for $y_S = 0.4$, $\beta_+(\beta_-) = 2.29(-2.31)$ at $r_h = 0.860(1.16)$ for $y_S = 0.6$, and $\beta_+(\beta_-) = 10.4(-10.3)$ at $r_h = 0.930(1.07)$ for $y_S = 0.8$.

This result demonstrates that the introduction of asymmetric electron distribution to open-shell singlet systems causes a remarkable enhancement of the maximum β amplitudes as compared to asymmetric closed-shell systems. On the other hand, a slight decrease in β is observed for large y_S in the region with large $|1 - r_h|$, and that a negligible β is also observed for r_h close to 1, where the sign of β changes. Overall, this implies that symmetric systems with large diradical characters can be chemically modified to display large or moderate β responses, either by introducing a sufficiently large asymmetricity or a moderate one.

4.6 Second Hyperpolarizability (γ) for Asymmetric Systems

In the same manner as that of β, we can derive a simple perturbative formula [Eq. (4.6.1)] for static γ along the bond axis (referred to as γ) in the three-state approximation {g, k, f} of the two-site diradical model [9, 36], where γ is partitioned into the contributions of types I(g1), II(g2), III-1(g31), and III-2(g32) (see Fig. 4.1a). Type I contribution is expressed as

$$\gamma^{\mathrm{I}} = 4\frac{(\mu_{\mathrm{kg}})^2(\Delta\mu_{\mathrm{kk}})^2}{(E_{\mathrm{kg}})^3} + 4\frac{(\mu_{\mathrm{fg}})^2(\Delta\mu_{\mathrm{ff}})^2}{(E_{\mathrm{fg}})^3}, \tag{4.6.1a}$$

where the first and second terms are referred to as g11(g-k-k-k-g) and g12(g-f-f-f-g), respectively. Type II contribution reads

$$\gamma^{\mathrm{II}} = -4\frac{(\mu_{\mathrm{kg}})^4}{(E_{\mathrm{kg}})^3} - 4\frac{(\mu_{\mathrm{fg}})^4}{(E_{\mathrm{fg}})^3} - 4\frac{(\mu_{\mathrm{kg}})^2(\mu_{\mathrm{fg}})^2}{E_{\mathrm{kg}}(E_{\mathrm{fg}})^2} - 4\frac{(\mu_{\mathrm{fg}})^2(\mu_{\mathrm{kg}})^2}{E_{\mathrm{fg}}(E_{\mathrm{kg}})^2}, \tag{4.6.1b}$$

where the first, second, third, and fourth terms are referred to as g21(g-k-g-k-g), g22(g-f-g-f-g), g23(g-k-g-f-g), and g24(g-f-g-k-g), respectively. Type III-1 contribution including the g311(g-k-k-f-g) (the first term) and g312(g-f-f-k-g) (the second term), and III-2 one including the g321(g-k-f-k-g) (the first term) and g322(g-f-k-f-g) (the second term) are expressed, respectively, as

$$\begin{aligned}\gamma^{\mathrm{III-1}} &= 8\frac{\mu_{\mathrm{gk}}\Delta\mu_{\mathrm{kk}}\mu_{\mathrm{kf}}\mu_{\mathrm{fg}}}{(E_{\mathrm{kg}})^2E_{\mathrm{fg}}} + 8\frac{\mu_{\mathrm{gf}}\Delta\mu_{\mathrm{ff}}\mu_{\mathrm{fk}}\mu_{\mathrm{kg}}}{(E_{\mathrm{fg}})^2E_{\mathrm{kg}}}, \text{ and}\\ \gamma^{\mathrm{III-2}} &= 4\frac{(\mu_{\mathrm{gk}})^2(\mu_{\mathrm{kf}})^2}{(E_{\mathrm{kg}})^2E_{\mathrm{fg}}} + 4\frac{(\mu_{\mathrm{gf}})^2(\mu_{\mathrm{fk}})^2}{(E_{\mathrm{fg}})^2E_{\mathrm{kg}}}.\end{aligned} \tag{4.6.1c}$$

As seen from these formulae, type I and III-1 contributions including dipole moment differences are only non-zero for molecular systems with asymmetric electron distribution, type I and III-2 contributions are positive, and type II terms are negative [6–8]. A nondimensional γ, defined by $\gamma/(R^4/U^3)$, is employed in the following discussion.

Figure 4.10 shows the y_S dependences of γ as well as of each contribution, type I(g1), II(g2), III-1(g31) and III-2(g32), for $r_h = 0.4$ (a), 0.8 (b) and 0.95 (c). Similar to the field application case [39], the total γ shows a bell-shape variation as a function of y_S, though the peak position (y_{Smax}) and amplitude ($\gamma_{\max}$) are different from each other: as increasing y_S, $\gamma_{\max}$ is significantly enhanced and y_{Smax} moves to higher y_S with r_h (<1) (see Fig. 4.11a). This behavior is similar to that observed for β (see Fig. 4.8a) as well as to the field dependence of γ for symmetric open-shell molecules [39]. This can be understood by the fact that the Hamiltonian matrix [Eq. (3.1.13)] for $(r_U, r_{tab}) = (1, 1)$ takes the same form as the Hamiltonian for a symmetric system under the effect of a static electric field F (Eq. (10) in Ref. [39]), where r_h replaces $r_F = FR/U$ [39]. Note that an electric field induced asymmetry is somewhat similar to an intra-molecular induced asymmetry, e.g., due to donor-acceptor substitution. Indeed, donor-acceptor substitutions lead to relatively local effects around the substituted positions whereas the field effects are

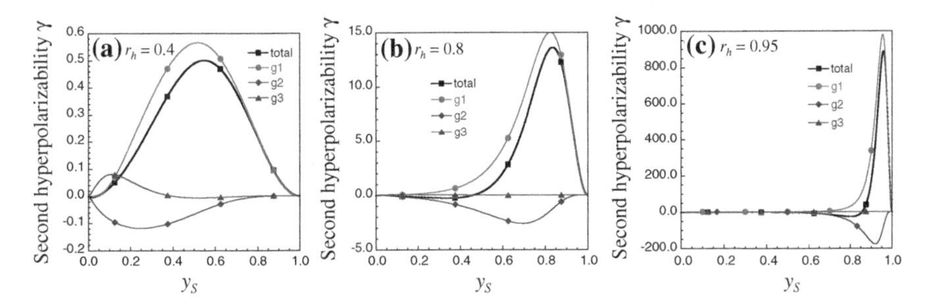

Fig. 4.10 y_S dependences of nondimensional γ [total, Type I(g1), Type II(g2), and Type III(g3)] for $r_h = 0.4$ (**a**), 0.8 (**b**), and 0.95 (**c**)

delocalized or more uniform, causing differences in the optical response properties in particular for extended conjugated molecules. As a result, for a given donor-acceptor pair, it has been shown that the same field amplitude cannot quantitatively reproduce the whole set of molecular response properties [38]. The primary contribution to γ originates from g1 (positive) for the whole y_S and r_h ranges, then from g2 (negative), which slightly shifts the maximum γ value towards larger y_S. Further analysis shows that g1 and g2 are dominated by g11(g-k-k-k-g) and g21(g-k-g-k-g), respectively. The important point is that the primary contribution originates from g11(g-k-k-k-g) in asymmetric diradical systems, while from g321(g-k-f-k-g) for symmetric ones [9]. A similar phenomenon has also been observed in symmetric two-site diradical models under static electric fields [39]. In real molecules, the field amplitudes achieving the maximum γ are often too strong to be realized, whereas the induction of asymmetric electron distribution (corresponding to a non-zero r_h) is considered as an alternative approach for realizing a sufficient asymmetricity and γ response. Furthermore, $r_U < 1 (r_U > 1)$, i.e., the electron affinity of A is larger(smaller) than B (see Sect. 3.1), is predicted to induce a positive(negative) contribution to the asymmetricity [through the off-diagonal elements, $(r_U - 1)/\sqrt{2}(r_U + 1)$, in Eq. (3.1.13)], though it has not been considered yet. As seen from Eq. (4.6.1a) and Fig. 3.3, the enhancement of g11 and the movement of y_{Smax} towards higher y_S values with r_h (<1) originate from the increase in $|\Delta\mu_{\mathrm{kk}}|/R$ and the decrease in E_{kg}/U with r_h (<1), as well as from the growing $|\mu_{\mathrm{gk}}|/R$ peak around $r_h = 1$ with y_S. The γ variation as a function of y_S for $r_h < 1$ and >1 are shown in Fig. 4.11a and b, respectively. It is found that $\gamma_{\max}$ (positive) increases and y_{Smax} moves towards higher y_S value with r_h (<1), while $\gamma_{\max}$ (positive) decreases and y_{Smax} moves towards smaller y_S value with r_h (>1), e.g., $(y_{\mathrm{Smax}}, \gamma_{\max}) = (0.548, 0.501)$ for $r_h = 0.4$, (0.759, 3.99) for $r_h = 0.7$, (0.836, 13.6) for $r_h = 0.8$, (0.834, 14.3) for $r_h = 1.2$, (0.753, 4.31) for $r_h = 1.3$, and (0.539, 0.562) for $r_h = 1.6$. Moreover, an emergence of negative γ in the intermediate y_S region is observed for r_h close to 1, e.g., negative $\gamma_{\max} = -0.250(-0.713)$ at $y_S = 0.339(0.404)$ for $r_h = 0.8(1.2)$. This occurs when |g11(positive)| < |g21(negative)| [Eqs. (4.6.1a) and (4.6.1b)]. The mirror symmetry of the r_h dependences of $\gamma_{\max}$ and y_{Smax} for $r_h > 1$ with respect to those for $r_h < 1$

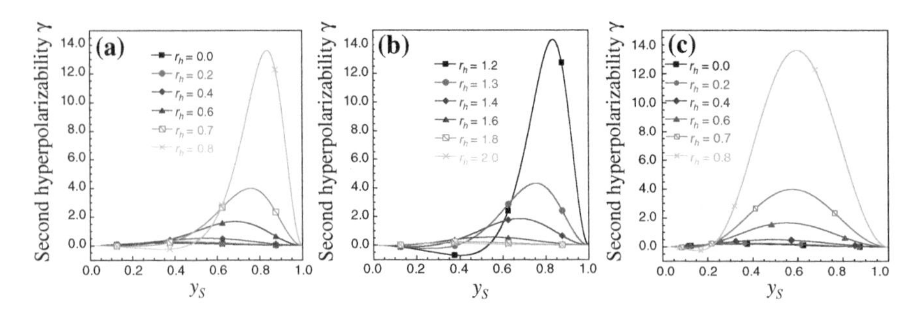

Fig. 4.11 Diradical character dependences of nondimensional γ as a function of r_h: y_S–γ for $r_h = 0.0 - 0.8$ (**a**) and $1.2 - 2.0$ (**b**), as well as $y_A - \gamma$ for $r_h = 0.0 - 0.9$ (**c**)

comes from the inversion in E_{kg}/U, $\Delta\mu_{kk}/R$, and $|\mu_{gk}|/R$ with respect to $r_h \sim 1$ for intermediate and large y_S (see Fig. 3.3e, f). In addition, as seen from the y_A–γ plot (Fig. 4.11f), for $r_h < 1$, γ is maximized for intermediate y_A values for any r_h, and γ_{max} increases with r_h, because of the relationship between y_A and y_S (Fig. 3.2).

The r_h dependences of γ for different y_S are shown in Fig. 4.12. All the γ variations are similar to each other except for their amplitudes and the r_h values giving local extrema of γ. As increasing r_h, the positive γ increases, attains a maximum (positive, referred to as γ_+), decreases toward a negative peak value (referred to as γ_-), the amplitude of which is larger than that of positive peak, and then it increases, becomes positive again, attains a maximum before tending to zero. When $y_S = 0$, the qualitative behavior is similar to that observed for closed-shell systems when switching an external field [37, 38, 40]. On the other hand, for

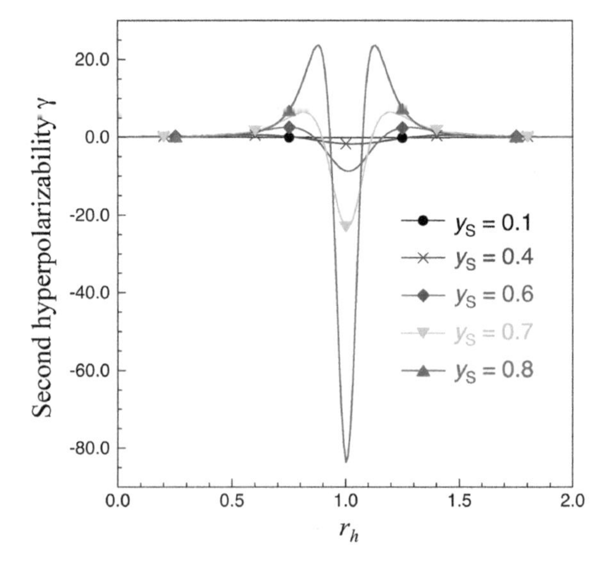

Fig. 4.12 r_h dependences of nondimensional γ for y_S ($0.1 \leq y_S \leq 0.8$)

two-site diradical systems, as increasing y_S, the amplitudes of the γ variations get larger when the γ positive peak gets closer to the position of the negative extremum ($r_h = 1$), e.g., $\gamma_+ = 0.526(0.515)$ at $r_h = 0.580(1.48)$ and $\gamma_- = -1.80$ at $r_h = 1.03$ for $y_S = 0.4$; $\gamma_+ = 6.51(6.52)$ at $r_h = 0.810(1.20)$ and $\gamma_- = -23.0$ at $r_h = 1.01$ for $y_S = 0.7$; and $\gamma_+ = 23.6(23.6)$ at $r_h = 0.880(1.13)$ and $\gamma_- = -83.5$ at $r_h = 1.00$ for $y_S = 0.8$. Negative γ occur when $|g11(\text{positive})| < |g21(\text{negative})|$ in Eqs. (4.6.1a) and (4.6.1b), which corresponds to $|\Delta\mu_{kk}| < |\mu_{gk}|$. As seen from Fig. 3.3d–f, this condition is satisfied by a rapid decrease of $|\Delta\mu_{kk}|/R$ toward zero and by a peak emergence of $|\mu_{gk}|/R$ around $r_h = 1$ for intermediate and large y_S. As a result, the introduction of asymmetric electron distribution to open-shell singlet molecular systems is predicted to cause remarkable enhancements (with positive or negative sign) of the local maximum γ amplitudes as compared to closed-shell asymmetric systems. On the other hand, it is predicted that a slight decrease in γ is observed in the case of large y_S in the region with large $|1 - r_h|$, and that a negligible γ is also observed for r_h close to 1, where the γ sign change occurs. However, large asymmetricity (r_h) regions can be realized by adjusting donor/acceptor substituents or by tuning the applied electric field intensity, giving the possibility to enhance γ in asymmetric open-shell singlet systems (with intermediate y_A) as compared to asymmetric closed-shell and symmetric open-shell singlet systems of similar size with intermediate y_S.

4.7 Second Hyperpolarizability (γ) for Multiradical Systems

In previous sections, we investigated diradical systems, which involve weakly interacting two unpaired electrons, and clarified the correlation between (hyper)polarizabilities and diradical character concerned with the occupation numbers of HONO and LUNO. On the other hand, such correlation is expected to be extended to singlet multiradical systems, which involve more than two unpaired electrons. Indeed, in recent studies on graphene nanoflakes, which exhibit multiradical character as increasing the size [41–43], the second hyperpolarizability (γ) is found to evolve remarkably [42] or non-monotonically [43] as a function of the size and γ displays a specific dependence on the multiple diradical characters y_i. In order to reveal these unique features, we here examine the γ values of linear H_4 chains, which are prototypical tetraradical systems, and the relationship between γ and multiple diradical characters y_i by varying the interatomic distances. The results are expected to contribute to the construction of molecular design guidelines for enhancing and controlling the third-order NLO properties of "singlet multiradical linear conjugated molecules".

4.7.1 Linear H_4 Model, Multiple Diradical Character, and Virtual Excitation Processes in the Perturbative Expression of γ

Figure 4.13 shows the symmetric linear H_4 models where the outer (r_1) and inner (r_2) H–H distances ranges from 1.0 to 4.0 Å, causing variations in the multiple diradical characters (y_0, y_1) between 0.0 (closed-shell) and 1.0 (pure diradical) [44]. The full configuration interaction (CI) method with a minimal basis set, STO-3G, is employed to calculate γ. Although such minimal basis set is known to be insufficient for providing quantitative γ values, it is sufficient for the present purpose because we obtain (i) almost the same bell-shape dependence of γ as a function of y_0 for the H_2 dissociation model as with a full CI method employing an extended basis set (aug-cc-pVDZ), and (ii) almost the same $\gamma_{max}(H_4)/\gamma_{max}(H_2)$ ratio and similar (y_0, y_1) values at these maxima [44].

The multiple diradical characters y_i are defined as the occupation numbers of the LUNO + i [45], where $i = 0, 1, \ldots$ (LUNO = Lowest Unoccupied Natural Orbital) calculated here using the full CI/STO-3G method:

$$y_i = n_{\mathrm{LUNO}+i}, \tag{4.7.1}$$

where y_i takes a value ranging from 0 to 1. In the H_2 model, y_0 (occupation number of the LUNO) increases from 0 to 1 with increasing the interatomic distance [46] which corresponds to the increase in the mixing of the anti-bonding (LUNO) and bonding [HONO (Highest Occupied Natural Orbital)] orbitals in the singlet ground-state wavefunction. Details of definitions and physical meanings of the diradical character are discussed in Sect. 2.2. The odd electron number ($2y_i$) and its density, which is defined by the ($n_{\mathrm{LUNO}+i}$-weighted LUNO + i density + $n_{\mathrm{HONO}-i}$-weighted HONO − i density), are not observable, but they provide the indices of chemical bond related to the (HONO − i, LUNO + i) pair. These measures are useful for obtaining intuitive and pictorial descriptions of the open-shell characters and of their impact on various response properties in multiradical systems [45–47].

The full CI/STO-3G method is applied to the calculation of the excitation energies and transition moments between the ground and excited states, which are then used to evaluate γ within the sum-over-states (SOS) formalism. All the calculations are performed using the GAMESS program package [48–50]. The full CI/STO-3G method provides model exact solutions, which avoid suffering from the incomplete treatment of static and dynamic electron correlations. For a

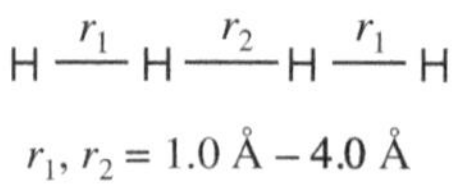

$$\mathrm{H} \overset{r_1}{—} \mathrm{H} \overset{r_2}{—} \mathrm{H} \overset{r_1}{—} \mathrm{H}$$

$$r_1, r_2 = 1.0\ \text{Å} - 4.0\ \text{Å}$$

Fig. 4.13 Sketch of the H_4 model linear hydrogen chain with interatomic distances r_1 and r_2

symmetric system, the perturbative SOS formula of the longitudinal static γ value [1, 2, 6–8] contains only type II (0–n–0–m–0) and type III-2 (0–n–m–n'–0) terms [see Eq. (4.2.1)], which correspond to virtual excitation processes shown in Fig. 4.13. Note that the sign of γ is determined by the balance between the negative type II and positive type III-2 contributions.

4.7.2 Effect of the Interatomic Distances on γ

The variations in γ(total), γ^{II}, and $\gamma^{\mathrm{III\text{-}2}}$ calculated at the SOS/full CI/STO-3G level in the r_1–r_2 plane are shown in Fig. 4.14 together with the schematic structures of the H_4 model at the four corners: (i) closed-shell H_4 with (r_1, r_2) [Å] = (1.0, 1.0), (ii) two independent closed-shell H_2 with (r_1, r_2) [Å] = (1.0, 4.0), (iii) both-end H atoms with a middle closed-shell H_2 with (r_1, r_2) [Å] = (4.0, 1.0), and (iv) four independent H atoms with (r_1, r_2) [Å] = (4.0, 4.0) [44]. The maximum γ (1,839 a.u.) is located at (r_1, r_2) [Å] = (1.60, 1.40) with (y_0, y_1) = (0.527, 0.178), which is enhanced by a factor of 115 as compared to twice the nearly closed-shell H_2 (H–H distance $r = 1.00$ Å) γ (8.01 a.u.). There are two channels of γ enhancement: one corresponds to a r_1 value close to 1.6 Å and describes an enhancement of γ when r_2 decreases from 4.0 to 1.40 Å, while the other one is parallel to the r_1 axis ($r_2 \sim 1.8$ Å) and shows an enhancement of γ with decreasing r_1 from 4.0 to 1.60 Å. These enhancements are caused by the type III-2 contributions [$\gamma_{\max}^{\mathrm{III-2}} = 2{,}478$ a.u. at (r_1, r_2) [Å] = (1.50, 1.30) with (y_0, y_1) = (0.447, 0.132)], while type II term is particularly important at small bond lengths, where $\gamma_{\min}^{\mathrm{II}} = -856$ a.u. at (r_1, r_2) [Å] = (1.30, 1.10) with (y_0, y_1) = (0.294, 0.067), and it modulates the type III-2 contribution and impacts the position of $\gamma_{\max}$.

4.7.3 Effect of the Diradical Characters on γ

The variations of γ as a function of the y_0 and y_1 ($y_0 \geq y_1$) diradical characters are shown in Fig. 4.15 [44]. It is found that γ attains the maximum at (y_0, y_1) = (0.527, 0.178), and that it decreases as going towards each side and apex of the triangle plane defined by the y_0 and y_1 axes. At the maximum ($\gamma_{\max} = 1{,}839$ a.u.), the primary contribution ($\gamma^{\mathrm{III-2}} = 2{,}404$ a.u.) comes from two dominant type III-2 virtual excitation processes, (0–2–3–2–0) (1,606 a.u.) and (0–2–4–2–0) (1,104 a.u.) (see Fig. 4.16), while γ^{II} (=−566 a.u.) mostly from (0–2–0–2–0) (−541 a.u.). The larger $\mu_{2,3}$ and $\mu_{2,4}$ transition moments with respect to $\mu_{0,2}$ together with the smaller energy gap between states 2 and 3(4) are in agreement with those of open-shell singlet systems with an intermediate diradical character, as shown in the two-site VCI model [9].

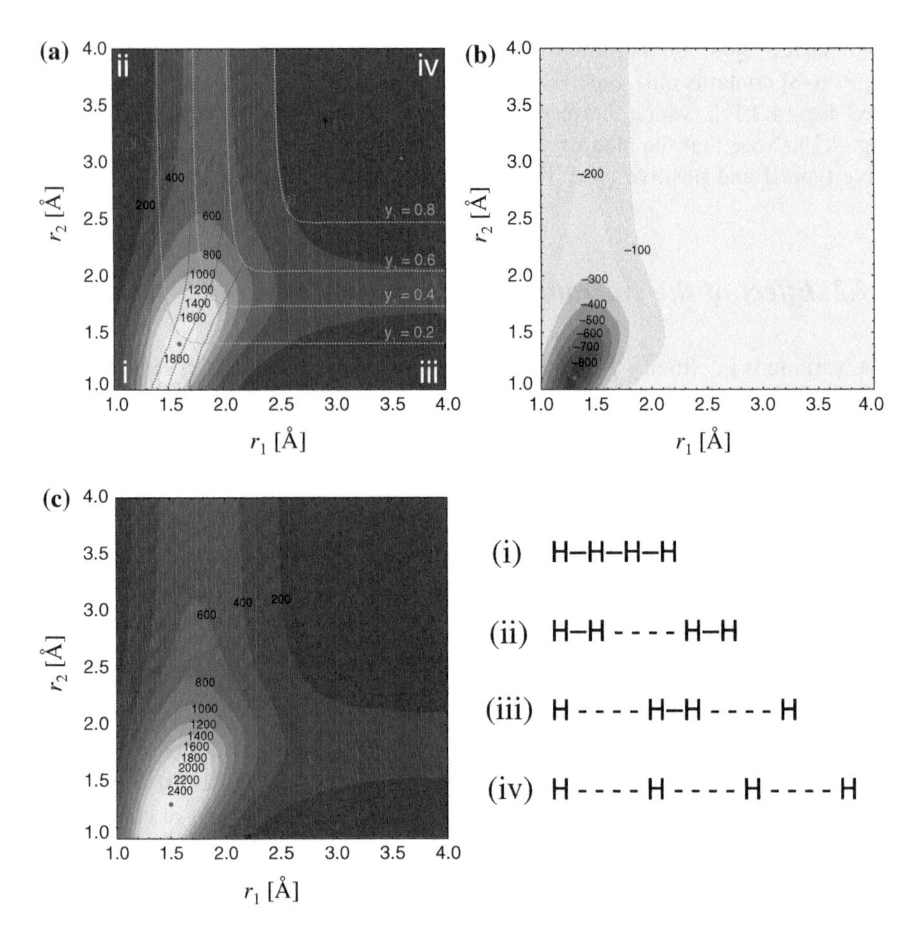

Fig. 4.14 Contours of γ values [total (**a**), type II (**b**) and type III-2 (**c**)] calculated using the full CI/STO-3G method in the r_1–r_2 plane as well as iso-y_0 (*red dotted*) and iso-y_1 (*green dotted*) *lines* in (**a**). *Red* points indicate the positions of the corresponding γ_{max}, (r_1, r_2) [Å] = (1.60, 1.40) (**a**), (1.30, 1.10) (**b**), and (1.50, 1.30) (**c**). Structures of H_4 models for the four corners (i)–(iv) are also shown

In order to clarify the relationship between (y_0, y_1) and (r_1, r_2), we can examine the variations in γ together with the iso-y_i lines (Fig. 4.14a). The enhanced γ zones satisfy the $0.2 \leq y_0$ and $y_1 \leq 0.8$ conditions, i.e., H_4 presents a tetraradical character whereas the maximum value lies in the narrow region with $0.5 \leq y_0 \leq 0.6$ and $0.1 \leq y_1 \leq 0.2$. The enhancement of γ is much smaller in the region with $y_0 \sim y_1$ even though both y_0 and y_1 lie in the intermediate region. This suggests that the γ values for one-dimensional symmetric singlet multiradical systems tend to be maximized for intermediate y_0 but smaller y_i $(i \geq 1)$ values, while $y_i \sim y_0$ situations lead to smaller γ. The origin of these enhancements can be found in the relationship between the diradical characters and the electronic structures of the H_4 system as follows. In general, y_i is correlated with the energy

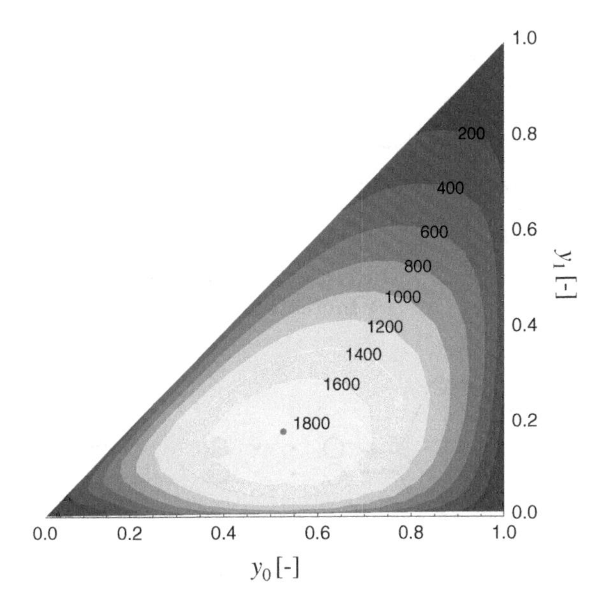

Fig. 4.15 Contours of γ in the y_0–y_1 plane ($y_0 \geq y_1$), where γ is maximized at (y_0, y_1) = (0.527, 0.178) (indicated by *red dot*)

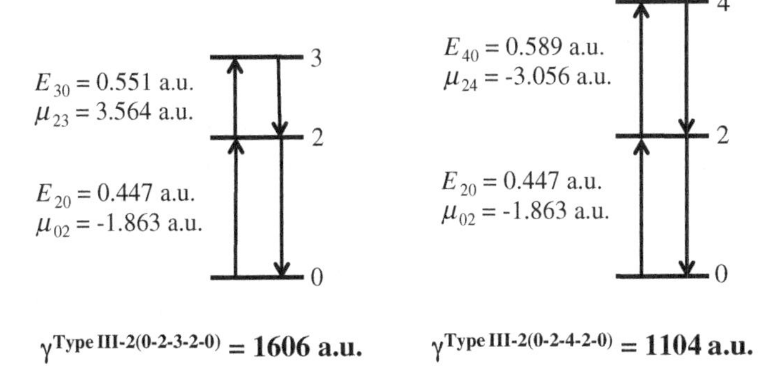

Fig. 4.16 Excitation energies E_{ij} and transition moments μ_{ij} of the primary type III-2 virtual excitation processes contributing to γ

gap between the highest occupied molecular orbital (HOMO) − i and the lowest unoccupied molecular orbital (LUMO) + i obtained using spin-restricted based methods like the spin-restricted Hartree-Fock (RHF) method since y_i is approximately defined by twice the weight of the corresponding doubly-excited configuration as shown in Sect. 2.2. Namely, the smaller(larger) the (HOMO − i) − (LUMO + i) gap the larger(smaller) the y_i value. We now examine the MO correlation diagrams in two cases: (a) when increasing r_1 for a fixed r_2, and (b) when increasing r_2 with a fixed r_1 (Fig. 4.17). The four frontier MOs of H_4,

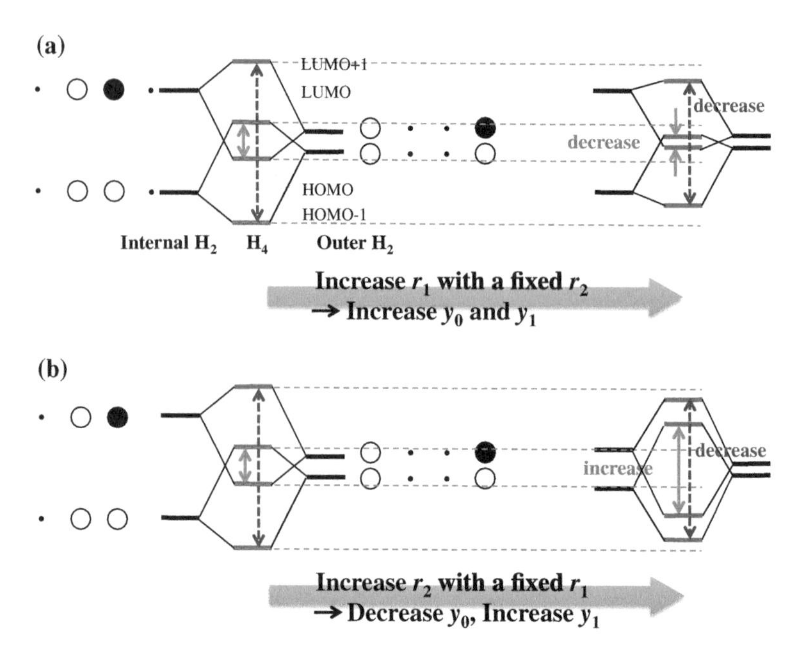

Fig. 4.17 Effects of r_1 (**a**) and r_2 (**b**) on the MO correlation diagram of H_4. The *left*- and *right-hand-side* levels (*solid bars*) represent the inner and outer two-site MOs (the phase of which is shown by *white* and *black circles*), while the center (*solid bars*) corresponds to the resulting four-site MO levels (HOMO − 1, HOMO, LUMO, LUMO + 1). The *solid* and *dotted* two-sided *arrows* point out the HOMO − LUMO and (HOMO − 1) − (LUMO + 1) gaps, respectively

in which the lower two are occupied, can be constructed by the correlation between the inner two-site MOs and the outer two-site MOs. When increasing r_1, the energy splitting between the outer MOs decreases whereas both energy splittings decrease when r_2 increases. Also, the energy splitting caused by the correlation between the outer and inner MOs is reduced with increasing r_1, while it is almost unchanged for a fixed r_1. As a consequence, in case (a), the HOMO/LUMO and (HOMO − 1)/(LUMO + 1) gaps decrease, leading to an increase of y_0 and y_1. In particular, in the limit of large r_1, y_0 is close to 1, while y_1 approaches the diradical character determined by the middle H_2 system. On the other hand, in case (b), the HOMO/LUMO gap increases while the (HOMO − 1)/(LUMO + 1) gap decreases, which corresponds to a decrease of y_0 and an increase of y_1. So, in the limit of large r_2, y_0 and y_1 become close to each other and are determined by the diradical character of each H_2 unit. This demonstrates that even in this case of intermediate diradical characters, the $y_0 \sim y_1$ condition corresponds to the situation where the left- and right-hand-side H_2 systems are nearly independent of each other. This is why 1D molecular aggregates, symmetric and singlet, where the two systems present similar intermediate y_i values, exhibit usually small γ values. On the other hand, to achieve a large (optimum) γ value, y_0 should belong to the intermediate region whereas y_1 be more than twice smaller, which corresponds to

$r_1 > r_2$ (Fig. 4.14a). Indeed, considering r_1 and $r_2 \in [1.0, 2.0]$ Å, when $r_1 > r_2$ the induced (third-order) polarization between both-end regions contributes more.

In this section, we employ the full CI method with a minimal basis set to the linear H_4 model (a prototype of one-dimensional symmetric singlet multiradical systems), and found that significant γ enhancements are achieved for intermediate y_0 and a small y_1 values, which corresponds to an inner H–H distance of 1.40 Å and an outer distance of 1.60 Å. In such a case, γ_{max} appears at $(y_0, y_1) = (0.527, 0.178)$ and is about 9 times larger than that of the nearly closed-shell H_4 ($y_0 = 0.099$, $y_1 = 0.029$ with all H–H distances of 1.0 Å). On the other hand, this enhancement is strongly reduced when $y_0 \sim y_1$. This analysis is useful for providing a molecular design guideline of highly efficient third-order NLO systems based on one-dimensional symmetric singlet multiradical systems, e.g., multiradical organic aggregates [15], supermolecular systems [51], and extended metal atom chains [52].

4.8 Examples of Open-Shell Singlet Molecular Systems

In this section, we present calculated results of several open-shell singlet molecular systems, which belong to intermediate diradical character region, as well as chemical modification guidelines for controlling diradical characters. As explained in previous sections, singlet diradical systems with intermediate diradical characters are found to exhibit a significant enhancement of γ values as compared to the closed-shell and pure diradical systems based on H_2 dissociation model [53, 54] and the two-site diradical model [9]. In previous studies, we firstly confirmed the validity of this structure-property relationship using several singlet diradical π-conjugated systems such as *p*-quinodimethane model (see Fig. 4.18) [55], twisted ethylene model [55], π-conjugated molecules involving imidazole and triazole rings [56], diphenalenyl diradical molecules [57], zethrenes [58] and so on (see Fig. 4.19). From these results, we can present a guideline for controlling the diradical character: (i) the systems have contributions of both benzenoid and quinoid resonance structures in the ground state, and (ii) the increase in the contribution of benzenoid (quinoid) structure tends to increase the diradical (closed-shell) character as seen from the resonance structure. The preference between benzenoid and quinoidal structures are expected to be controlled by tuning the molecular architectures and/or by introducing donor/acceptor substitutions. As easily predicted, the increase (decrease) in the aromaticity in the middle ring region of fused-ring systems such as diphenalenyl diradicaloids corresponds to the increase (decrease) in the diradical character. Namely, the increase in the size of acene region in the molecules like IDPL (see Fig. 4.19a and d) and the substitution of an atom in the central ring by that with high electronegativity of the molecules like TDPL (see Fig. 4.19c) are expected to increase the diradical character. The introduction of acceptor substituents into the both end of quinoidal oligothiophenes are predicted to increase the diradical character by localizing the two radicals

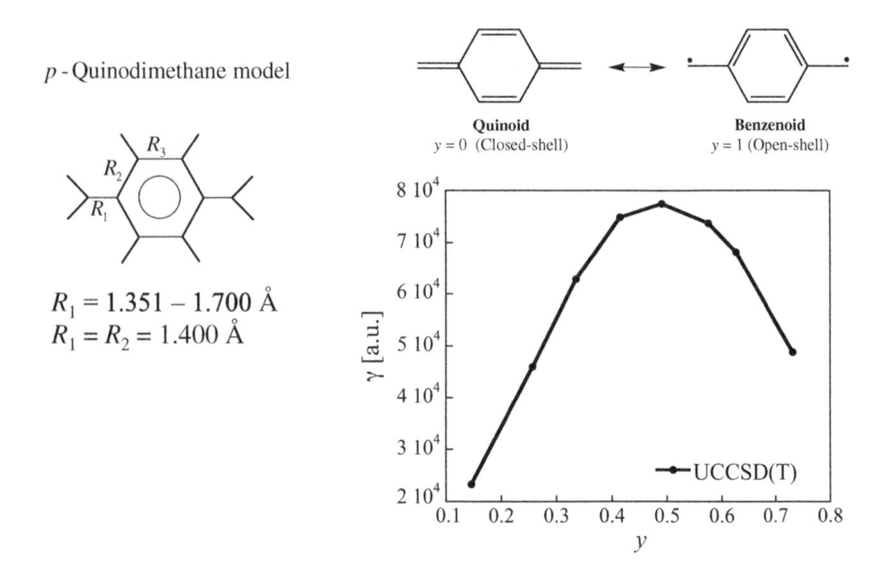

Fig. 4.18 *p*-Quinodimethane model and the diradical character (y) dependence of γ at the UCCSD(T) level of approximation. Resonance structures, quinoid (closed-shell) and benzenoid (open-shell) forms, are also shown

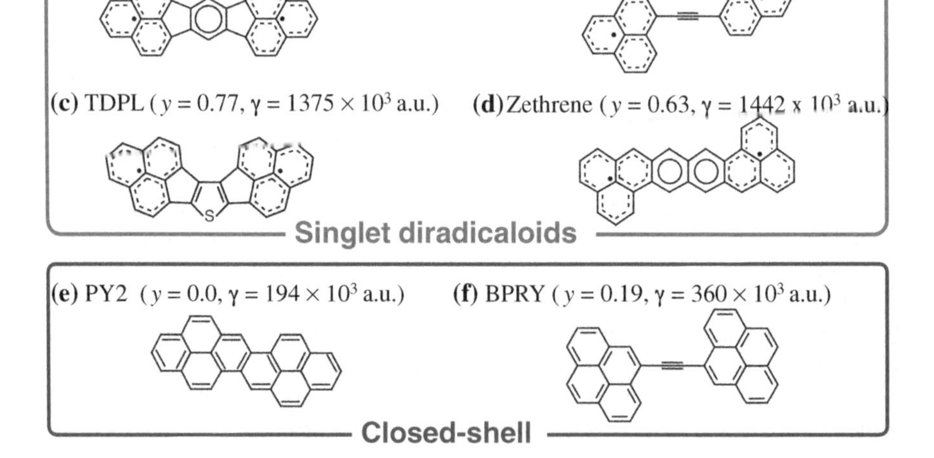

Fig. 4.19 Examples of singlet diradicaloids including phenalenyl rings and closed-shell analogs. Diradical characters (y) (at the PUHF level of approximation) and γ values (at the UBHHLYP level of approximation) are also shown

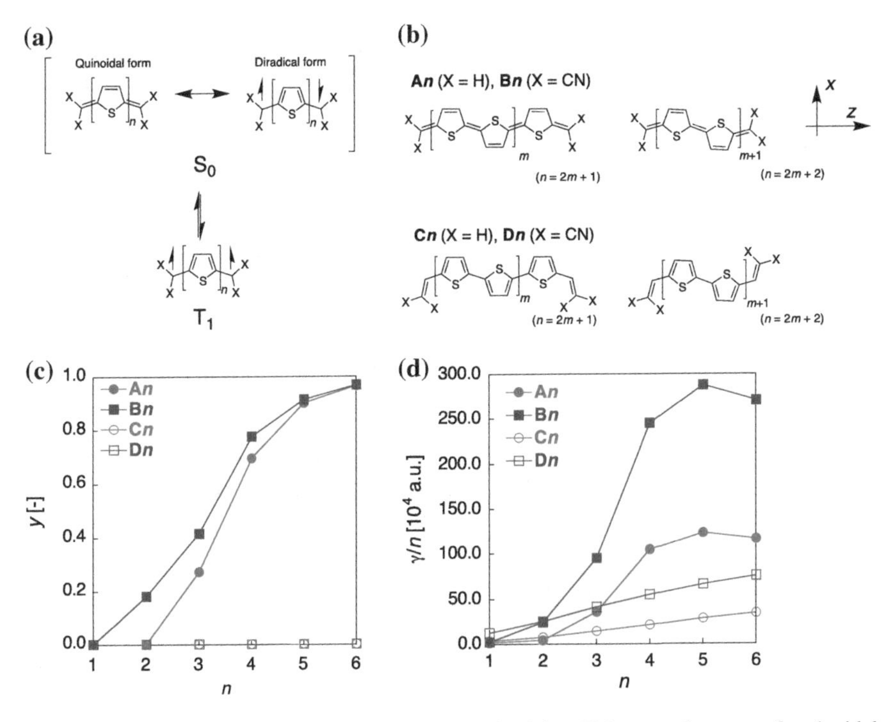

Fig. 4.20 Molecular structures of singlet (S_0) and triplet (T_1) ground states of quinoidal oligothiophenes (QTs) (**a**), and calculated models: open-shell systems **A*n*** (X = H) and **B*n*** (X = CN) and closed-shell systems **C*n*** (X = H) and **D*n*** (X = CN) for n = 1–6 (m = 0, 1, and 2). Coordinate axes are also shown. Chain-length (n) dependences of diradical characters (y) (**c**) and γ/n (**d**) for systems **A*n***, **B*n***, **C*n***, and **D*n***

on the both-end carbon (C) atoms, respectively, and to decrease the contribution of quinoidal form (see Fig. 4.20). The tuning of the size of the middle thiophene ring and by changing the substitution form at the both ends are predicted to significantly change the diradical character [59].

The next candidates for open-shell singlet molecules are "graphene nanoflakes (GNFs)" [60–65], which are finite size graphenes with several architectures, e.g., trigonal, rectangular, hexagonal and antidot structures, and unique edge structures, i.e., zigzag and armchair edges. The open-shell singlet nature of finite-size GNFs is caused by the near-degenerate the highest occupied molecular orbital (HOMO) and the lowest unoccupied molecular orbital (LUMO) [66–68]. This is alternatively understood by their resonance structures based on Clar's sextet rule [69], which states that resonance structures having a larger number of six-membered rings with benzenoid structures (Clar's sextet) exhibit larger aromatic stabilities and tend to make more significant contributions to their ground electronic states. Figure 4.21a shows an example of pentacene, which is regarded as one of rectangular GNFs and the open-shell resonance form with radicals located on the middle region of the zigzag edges shows more Clar's sextet than the closed-shell

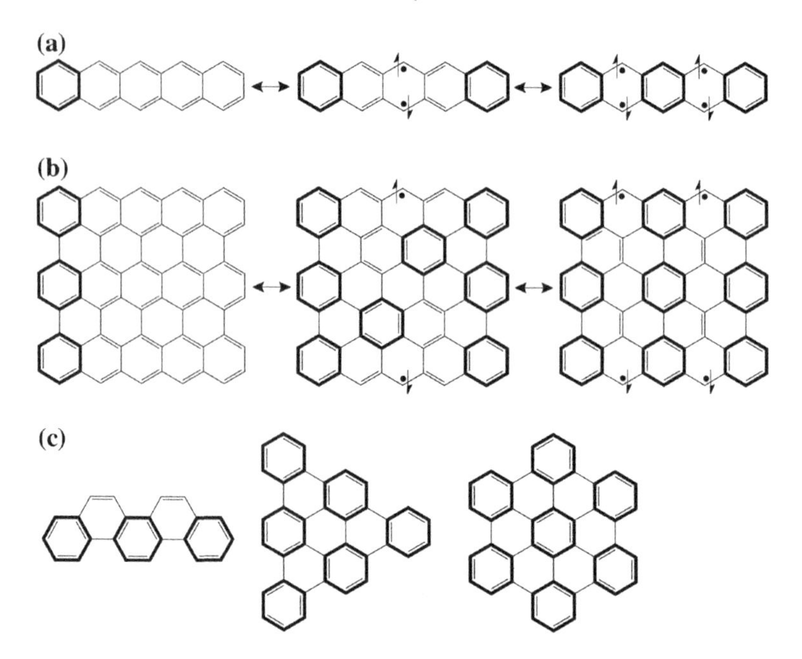

Fig. 4.21 Resonance structures for pentacene (**a**) and a larger rectangular GNF (**b**), and linear, trigonal, and hexagonal GNFs having only armchair edges (**c**)

form. The similar feature is also observed for larger GNFs as shown in Fig. 4.21b. On the other hand, for GNFs having only armchair edges, their closed-shell resonance forms show maximal numbers of Clar's sextet regardless of their molecular architectures including linear, trigonal, and hexagonal forms (see Fig. 4.21c). As a result, the GNFs having zigzag edges tend to possess open-shell singlet ground states, where primary spin polarization is observed between the both-end zigzag edges, while those having only armchair edges tend to possess closed-shell ground states. Let us consider the edge effect on the γ components of rectangular GNFs [41]. We examine the γ_{xxxx} and γ_{yyyy} values, which are the components along its zigzag and armchair edges, respectively, of PAH[3,3] with intermediate diradical character in the singlet and triplet states calculated at the UBHHLYP/6-31G* level of approximation. For PAH[3,3] in the singlet state, the γ_{yyyy} (14.45×10^4 a.u.) is about four times as large as γ_{yyyy} (3.412×10^4 a.u.). Because γ_{iiii} is generally expected to increase nonlinearly with increasing the π-conjugation length in the i-direction, the significant deference between γ_{xxxx} and γ_{yyyy} for singlet PAH[3,3] despite of its square molecular structure is not usual. The exceptional enhancement of γ_{yyyy} is predicted to stem from the intermediate y_0 value (0.510). In general, the γ enhancement effect originating from intermediate diradical character is observed in the component of direction from one radical site to the other radical site, which corresponds the direction from one zigzag edge to the opposite one, i.e., y-direction, in the case of rectangular GNFs. There is another evidence of the open-shell NLO systems for singlet PAH[3,3], i.e., significant spin state dependence of γ_{yyyy}. It is predicted that the intermediate diradical systems exhibit a significant reduction of γ amplitude by going from the singlet to the triplet state due to the Pauli effect in the triplet state [70, 71]. Indeed, the γ_{yyyy} value exhibits a significant reduction by changing from the singlet ($\gamma_{yyyy} = 14.45 \times 10^4$ a.u.) to triplet state ($\gamma_{yyyy} = 3.501 \times 10^4$ a.u.), while there is a little difference in the γ_{xxxx} value between the singlet ($\gamma_{xxxx} = 3.412 \times 10^4$ a.u.) and triplet states ($\gamma_{xxxx} = 3.412 \times 10^4$ a.u.). As a result, the large γ_{yyyy} for singlet PAH[3,3] is found to be caused by the intermediate diradical character. Larger size PAHs[*X*,*Y*] are found to exhibit multi-radical effects on the γ and its size dependences, where the γ values are also found to be enhanced in the region with intermediate multiple diradical characters y_i (concerning the HOMO $- i$ and LUMO $+ i$) [42]. Furthermore, the architectures and size of GNFs are found to significantly affect the diradical characters and then the γ values, the feature of which contribute to building the practical design guidelines for highly efficient open-shell NLO materials [72–76].

The third candidates are transition metal involving systems, in particular, metal–metal bonded systems [52]. It is known that in transition metal complexes, d–d orbital interactions often lead to multiple bonds of dσ, dπ, and dδ characters. These multiple bonds exhibit diradical characters from spin-unrestricted Hartree–Fock (UHF) and DFT calculations of the occupation numbers of the dσ, dπ, and dδ natural orbitals in the naked dichromium(II) model system [77, 78]. Also, the effective bond orders (EBO) of such systems are predicted to be generally smaller than their formal bond orders; for example, the $[Cr_2(O_2CCH_3)_4]$ dichromium(II) complex has a formal bond order of 4 but its EBO is 1.99 [79]. These results indicate characteristic weak metal–metal bonds and the possible emergence of singlet multiradical character. Therefore, transition metal complexes with multiple metal–metal bonds appear as promising systems for their third-order NLO properties. In our previous study [80], we have investigated the γ values of the singlet dichromium(II) [Cr(II)–Cr(II)] and dimolybdenum(II) [Mo(II)–Mo(II)] model systems along with the diradical characters of the dσ, dπ, and dδ orbitals and the contributions of dσ, dπ, and dδ electrons to γ [γ(dσ), γ(dπ), and γ(dδ), respectively], by varying the metal–metal bond lengths. It is found that γ(dσ) is dominant and takes a maximum value [γ_{max}(dσ)] in the intermediate diradical character regions of the dσ orbitals in both model systems, which indicates that open-shell singlet metal–metal bonded systems belong to a novel class of "σ-dominant" third-order NLO systems. As seen from Eq. (4.2.3), γ is also strongly affected by the diradical distance R_{BA}: γ is proportional to the fourth power of R_{BA}. Therefore, we found that an intermediate diradical character with a long A–B bond length leads to large γ_{max} value. This predicts that an intermediate dσ diradical character with a long metal–metal bond length leads to large γ_{max}(dσ) in metal–metal bonded systems, resulting in enhancement of γ_{max}. Because diffuse atomic orbitals can interact with each other at long distance, which leads to an intermediate diradical character with large bond length, metal–metal bonded systems composed of transition metals with a diffuse valence d atomic orbital are expected to satisfy the condition for enhancing γ_{max}. Therefore, the γ_{max} value of a metal–metal bonded system is predicted to depend on the group number, on the period number, and on the charge of metal atoms because the size of the valence d atomic orbital is related to them. It is found that smaller group number, larger period number and smaller positive charge of transition metals generally lead to larger size of the valence d atomic orbitals. Indeed, it is found that a dimetallic system composed of transition metals with smaller group number, larger period number and/or smaller positive charge will present an intermediate dσ diradical character along with a longer bond length and will thus exhibit larger γ_{max} (see, for example, Tables 4.2 and 4.3) [81]. Similar structure-property relationships are also observed in extended metal atomic chains [82]. The equatorial and axial ligand effects on the open-shell character and γ values are also investigated, and we found that the fundamental relationship between dσ interactions and γ is preserved and that the ligand effects can be used for tuning the diradical character, bond length and then further enhancing the amplitudes of γ_{max} [83]. These results provide a guideline for an effective molecular design of highly efficient third-order NLO systems based on the metal–metal bonded systems.

As other candidates, we investigate the molecules involving main group elements [84], which are known to have relatively weak bonds originating in the weak hybridization feature of heavy main group elements as compared to the second period atoms like B, C, N, and O atoms. Recent developments of synthesis

Table 4.2 Maximum UCCSD γ(dσ) [γ_{max}(dσ)], PUHF diradical character [y_{max}(dσ)] and bond length [R_{max}(dσ)] corresponding to γ_{max}(dσ) for V(II)–V(II), Cr(II)–Cr(II) and Mn(II)–Mn(II) as well as their γ_{max} and R_{max} obtained by the UCCSD and UCCSD(T) methods

	V(II)–V(II)	Cr(II)–Cr(II)	Mn(II)–Mn(II)
γ_{max}(dσ) [a.u.]	4,600	1,570	535
y_{max}(dσ) [−]	0.744	0.776	0.797
R_{max}(dσ) [Å]	3.0	2.8	2.6
γ_{max} [a.u.]	4,390 (4,480)[a]	1,570 (1,650)[a]	568 (599)[a]
R_{max} [Å]	3.1 (3.1)[a]	2.8 (2.8)[a]	2.6 (2.6)[a]

[a] Values in parentheses correspond to UCCSD(T) results

Table 4.3 Maximum UCCSD γ(dσ) [γ_{max}(dσ)], PUHF diradical character [y_{max}(dσ)] and bond length [R_{max}(dσ)] corresponding to γ_{max}(dσ) for Cr(II)–Cr(II), Mo(II)–Mo(II) and Mn(II)–Mn(II) as well as their γ_{max} and R_{max} obtained by the UCCSD and UCCSD(T) methods

	Cr(II)–Cr(II)	Mo(II)–Mo(II)	W(II)–W(II)
γ_{max}(dσ) [a.u.]	1,570	7,800	17,400
y_{max}(dσ) [−]	0.776	0.662	0.776
R_{max}(dσ) [Å]	2.8	3.4	4.1
γ_{max} [a.u.]	1,570 (1,650)[a]	7,630 (8,000)[a]	16,500 (17,600)[a]
R_{max} [Å]	2.8 (2.8)[a]	3.4 (3.4)[a]	4.1 (4.1)[a]

[a] Values in parentheses correspond to UCCSD(T) results

in the thermally stable molecular systems involving such main group elements are expected to open a way to realize novel structure–property relationships in NLO properties and to provide switching/controlling strategies of such properties. As another extension direction of open-shell functional molecular systems, super/supra-molecular architectures, e.g., multiradical aggregates, are promising. We have performed preliminary studies on the size and multiple diradical character dependences of NLO properties of model supra-/super-molecular open-shell systems, e.g., hydrogen linear chains with different bond length alternations as well as charges [85, 86], slipped-stack aggregates composed of diphenalenyl diradicaloids [87], one-dimensional fused-ring chain systems [43, 74]. These results suggest that the size dependences of the NLO properties in the relatively small multiple diradical character regions are more enhanced than the corresponding closed-shell multiradicaloids and/or their highest spin states [85]. Spin multiplicity effects, in particular, in the intermediate spin states and/or in different charge states on these multiradicaloids will provide further possible control schemes giving drastic changes in the NLO properties together with intriguing cooperative effects between optical and magnetic properties. Furthermore, such structure-property relationship on γ for symmetric molecules is predicted to be applied to the γ and β for asymmetric molecules using donor and acceptor disubstituted diphenalenyl radical compounds [88], where additional terms (involving dipole moment differences) originating asymmetric charge distributions contribute to β and γ as well as the contributions enhanced by the intermediate diradical character.

On the basis of our previous theoretical predictions, Kamada et al. have measured the two-photon absorption (TPA) properties of singlet diradical hydrocarbons involving diphenalenyl radicals, IDPL and NDPL, synthesized by Kubo et al. [12–15], and have found that the TPA cross section of NDPL records over 8,000 GM at 1,055 nm, which is the largest value reported so far for pure hydrocarbons [89]. Recently, other experimental verifications of our theoretical results have been performed for the TPA of benzannulated Chichibabin's hydrocarbons [90], hepta-/octa-zethrenes [25], and bis(acridine) dimers [91] as well as on the third harmonic generation of 1,4-bis-(4,5-diphenylimidazole-2-ylidene)-cyclohexa-2,5-diene (BDPI-2Y) [92]. From another point of view, these studies have clarified the diradical character dependences of the excitation energies and excitation properties of low-lying singlet excited states of open-shell molecules as well as of the singlet-triplet energy gap, which indicates that other intriguing optical and magnetic properties such as singlet fission [93], which is the main theme in Chap. 5, and half-metallicity [94] are also closely related to the diradical characters. In summary, the "open-shell character", i.e., "multiple diradical character" is considered to be a useful chemical concept, which governs the structure, property, and reactivity in a wide range of electron-correlated molecular systems, and thus presents novel design guidelines of molecular-based functional materials.

References

1. B.J. Orr, J.F. Ward, Mol. Phys. **20**, 513 (1971)
2. D.M. Bishop, J. Chem. Phys. **100**, 6535 (1994)
3. A. Willetts, J.E. Rice, D.M. Burland, D.P. Shelton, J. Chem. Phys. **97**, 7590 (1992)
4. P. Norman, K. Ruud, in *Non-Linear Optical Properties of Matter. From Molecules to Condensed Phases*, ed. by M.G. Papadpoulos, A.J. Sadlej, J. Leszczynsli (Springer, Netherland, 2006) pp. 1–49
5. D.M. Burland, in *Optical Non-Linearities in Chemistry*, ed. by D.M. Burland (American Chemical Society, Washington, D.C., 1994); special issue on optical nonlinearities in chemistry. Chem. Rev. **94**, 1–278 (1994)
6. M. Nakano, M. Okumura, K. Yamaguchi, T. Fueno, Mol. Cryst. Liq. Cryst. **182A**, 1 (1990)
7. M. Nakano, K. Yamaguchi, Chem. Phys. Lett. **206**, 285 (1993)
8. M. Nakano, I. Shigemoto, S. Yamada, K. Yamaguchi, J. Chem. Phys. **103**, 4175 (1995)
9. M. Nakano et al., Phys. Rev. Lett. **99**, 033001 (2007)
10. W. Heisenberg, Z. Phys. **49**, 619 (1928)
11. C. Lambert, Angew. Chem. Int. Ed. **50**, 1756 (2011)
12. T. Kubo et al., Angew. Chem. Int. Ed. **43**, 7474 (2004)
13. T. Kubo et al., Angew. Chem. Int. Ed. **44**, 6564 (2005)
14. A. Shimizu et al., Angew. Chem. Int. Ed. **48**, 5482 (2009)
15. A. Shimizu et al., J. Am. Chem. Soc. **132**, 14421 (2010)
16. A. Shimizu et al., Chem. Commun. **48**, 5629 (2012)
17. A. Konishi et al., J. Am. Chem. Soc. **135**, 1430 (2013)
18. A. Shimizu et al., Angew. Chem. Int. Ed. **52**, 6076 (2013)
19. D. Hibi et al., Org. Biomol. Chem. **11**, 8256 (2013)
20. Z. Sun et al., J. Chem. Soc. Rev. **41**, 7857 (2012)
21. R. Umeda et al., Org. Lett. **11**, 4104 (2009)
22. T.C. Wu et al., Angew. Chem. Int. Ed. **49**, 7059 (2010)
23. Z. Sun et al., Org. Lett. **12**, 4690 (2010)
24. Z. Sun et al., J. Am. Chem. Soc. **133**, 11896 (2011)
25. Y. Li et al., J. Am. Chem. Soc. **134**, 14913 (2012)
26. Z. Zeng et al., J. Am. Chem. Soc. **135**, 6363 (2013)
27. M. Nakano et al., AIP Conf. Proc. **1504**, 136 (2012)
28. R. Kishi, M. Nakano, J. Phys. Chem. A **115**, 3565 (2011)
29. T. Minami, S. Ito, M. Nakano, J. Phys. Chem. A **117**, 2000 (2013)
30. K. Kamada, K. Ohta et al., Angew. Chem. Int. Ed. **46**, 3544 (2007)
31. K. Ohta, K. Kamada, J. Chem. Phys. **124**, 124303 (2006)
32. M. Nakano et al., J. Chem. Phys. **131**, 114316 (2009)
33. K. Kamada, K. Ohta, Y. Iwase, K. Kondo, Chem. Phys. Lett. **372**, 386 (2003)
34. J. Perez-Moreno, K. Clays, M.G. Kuzyk, J. Chem. Phys. **128**, 084109 (2008)
35. W.-H. Lee, M. Cho, S.-J. Jeon, B.R. Cho, J. Phys. Chem. A **104**, 11033 (2000)
36. M. Nakano, B. Champagne, J. Chem. Phys. **138**, 244306 (2013)
37. F. Meyers, J.L. Brédas, J. Zyss, J. Am. Chem. Soc. **114**, 2914 (1992)
38. B. Kirtman, B. Champagne, D.M. Bishop, J. Am. Chem. Soc. **122**, 8007 (2000)
39. M. Nakano, B. Champagne et al., J. Chem. Phys. **133**, 154302 (2010)
40. V. Keshari, S.P. Karna, P.N. Prasad, J. Phys. Chem. **97**, 3525 (1993)
41. M. Nakano, H. Nagai et al., Chem. Phys. Lett. **467**, 120 (2008)
42. H. Nagai, M. Nakano, Chem. Phys. Lett. **489**, 212 (2010)
43. K. Yoneda, M. Nakano et al., ChemPhysChem **12**, 1697 (2011)
44. M. Nakano, T. Minami et al., J. Chem. Phys. **136**, 0243151 (2012)
45. M. Nakano, H. Fukui et al., Theoret. Chem. Acc. **130**, 711 (2011); erratum **130**, 725 (2011)
46. K. Yamaguchi, in *Self-Consistent Field: Theory and Applications*, ed. by R. Carbo, M. Klobukowski (Elsevier, Amsterdam, 1990), p. 727

47. K. Kamada, K. Ohta et al., J. Phys. Chem. Lett. **1**, 937 (2010)
48. M.W. Schmidt et al., J. Comput. Chem. **14**, 1347 (1993)
49. M.S. Gordon, M.W. Schmidt, in *Theory and Applications of Computational Chemistry, the first forty years*, ed. by C.E. Dykstra et al (Elsevier, Amsterdam, 2005)
50. Z. Gan et al., J. Chem. Phys. **119**, 47 (2003)
51. M. Nakano et al., J. Phys. Chem. A **115**, 8767 (2011)
52. F.A. Cotton, C.A. Murillo, R.A. Walton, *Multiple Bonds Between Metal Atoms*, ed. by F.A. Cotton, C.A. Murillo, R.A. Walton, 3rd edn. (Springer, New York, 2005)
53. M. Nakano et al., Phys. Rev. A **55**, 1503 (1997)
54. M. Nakano et al., J. Chem. Phys. **125**, 074113 (2006)
55. M. Nakano et al., J. Phys. Chem. A **109**, 885 (2005)
56. M. Nakano et al., J. Phys. Chem. A **110**, 4238 (2006)
57. M. Nakano, T. Kubo et al., Chem. Phys. Lett. **418**, 142 (2006)
58. M. Nakano et al., Comput. Lett. **3**, 333 (2007)
59. R. Kishi et al., J. Phys. Chem. C **117**, 21498 (2013)
60. W.L. Wang et al., Nano Lett. **8**, 241 (2008)
61. O.V. Yazyev et al., Nano Lett. **8**, 766 (2008)
62. M. Ezawa, Physica E **40**, 1421 (2008)
63. M. Ezawa, Phys. Rev. B **76**, 245415 (2007)
64. J.F.-Rossier et al., Phys. Rev. Lett. **99**, 177204 (2007)
65. J.R. Dias, Chem. Phys. Lett. **467**, 200 (2008)
66. M. Fujita, K. Wakabayashi, K. Nakada, K. Kusakabe, J. Phys. Soc. Jpn. **65**, 1920 (1996)
67. M. Bendikov et al., J. Am. Chem. Soc. **126**, 7416 (2004)
68. J. Hachmann, J.J. Dorando et al., J. Chem. Phys. **127**, 134309 (2007)
69. E. Clar, *Polycyclic Hydrocarbons* (Academic Press, London, 1964)
70. M. Nakano et al., J. Phys. Chem. A **108**, 4105 (2004)
71. S. Ohta et al., J. Phys. Chem. A **111**, 3633 (2007)
72. H. Nagai et al., Chem. Phys. Lett. **477**, 355 (2009)
73. K. Yoneda et al., Chem. Phys. Lett. **480**, 278 (2009)
74. S. Motomura et al., Phys. Chem. Chem. Phys. **13**, 20575 (2011)
75. K. Yoneda et al., J. Phys. Chem. C **116**, 17787 (2012)
76. K. Yoneda et al., J. Phys. Chem. Lett. **3**, 3338 (2012)
77. M. Nishino et al., J. Phys. Chem. A **101**, 705 (1997)
78. M. Nishino et al., Bull. Chem. Soc. Jpn **71**, 99 (1998)
79. B.O. Roos et al., Angew. Chem. Int. Ed. **46**, 1469 (2007)
80. H. Fukui et al., J. Phys. Chem. Lett. **2**, 2063 (2011)
81. H. Fukui et al., J. Phys. Chem. A **116**, 5501 (2012)
82. H. Fukui et al., Chem. Phys. Lett. **527**, 11 (2012)
83. Y. Inoue et al., Chem. Phys. Lett. **570**, 75 (2013)
84. F. Breher, Coord. Chem. Rev. **251**, 1007 (2007)
85. M. Nakano et al., Chem. Phys. Lett. **432**, 473 (2006)
86. A. Takebe et al., Chem. Phys. Lett. **451**, 111 (2008)
87. A. Takebe et al., Chem. Phys. Lett. **451**, 111 (2008)
88. M. Nakano et al., J. Phys. Chem. Lett. **2**, 1094 (2011)
89. K. Kamada et al., Angew. Chem. Int. Ed. **46**, 3544 (2007)
90. Z. Zeng et al., J. Am. Chem. Soc. **134**, 14513 (2012)
91. K. Kamada et al., J. Am. Chem. Soc. **135**, 232 (2013)
92. H. Kishida et al., Thin Solid Films **519**, 1028 (2010)
93. M.J. Smith, J. Michl, Chem. Rev. **110**, 6891 (2010)
94. Y.-W. Son, M.L. Cohen, S.G. Louie, Nature **444**, 347 (2006)

Chapter 5
Diradical Character View of Singlet Fission

Abstract Singlet fission is one of the internal conversion process in which a singlet exciton splits into two triplet excitons having long lifetimes. This phenomenon is expected to be useful for significantly improving the photoelectric conversion efficiency in organic photovoltaic cells. In this chapter, we present diradical character based molecular design guidelines for efficient singlet fission molecules based on the energy level matching conditions between the lowest singlet and triplet excited states, which are found to be described by the multiple diradical characters. A simple model, i.e., tetraradical hydrogen cluster, is investigated in order to reveal the multiple diradical character dependences of relative excitation energies and to build a diradical character based design guideline. On the basis of this guideline, several candidate molecules are proposed.

Keywords Singlet fission · Triplet exciton · Energy level matching · Photoelectric conversion · Diradical character · Tetraradical character

5.1 Singlet Fission

5.1.1 Singlet Fission from the Viewpoint of Photoelectric Conversion

Electronic excitations are one of the fundamental topics in chemistry, physics, biology and materials science since they play an important role in optical absorptions, reflections, emissions, magnetism and so on. These phenomena are described by a concept "a quasi-particle", i.e., "exciton" [1, 2], which is defined by a pair of excited electron and hole bounded by Coulomb interaction originating in their mutually opposite charges (Fig. 5.1a). Using this concept, an optical absorption, emission, and spatial propagation of excitations are interpreted as a generation, recombination, and spatial transfer of exciton, respectively. This chapter concerns the singlet fission (SF) (Fig. 5.1b) [3] of exciton in organic

M. Nakano, *Excitation Energies and Properties of Open-Shell Singlet Molecules*,
SpringerBriefs in Electrical and Magnetic Properties of Atoms, Molecules, and Clusters,
DOI 10.1007/978-3-319-08120-5_5

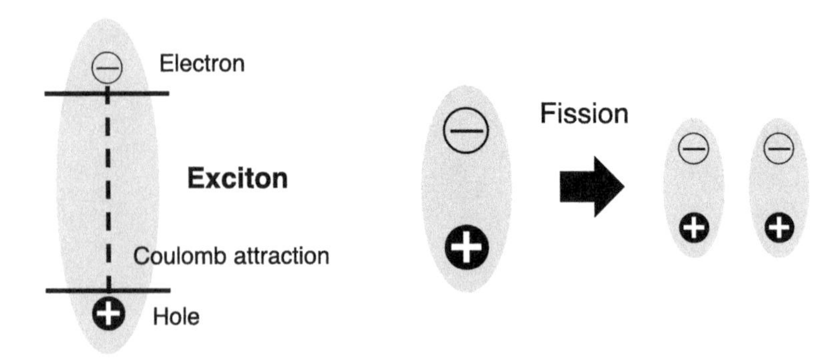

Fig. 5.1 Schematic picture of exciton (**a**) and singlet fission (**b**)

molecular systems, where a single singlet exciton splits into two triplet excitons. Singlet fission is also regarded as one of the reverse processes of triplet annihilation, where two triplet excitons collide and fuse into a singlet exciton. Although singlet fission was first observed in 1965 for anthracene crystal [4], only a little attention has been paid until Nozik and co-workers have emphasized its importance in the application to photovoltaics [5].

The process of singlet fission is approximately described by

$$S_1 + S_0 \underset{k_2}{\overset{k_{-2}}{\rightleftarrows}} {}^1(\mathrm{T_1T_1}) \underset{k_1}{\overset{k_{-1}}{\rightleftarrows}} \mathrm{T_1} + \mathrm{T_1}. \tag{5.1.1}$$

Here, $^1(\mathrm{T_1T_1})$ is the singlet state composed of two triplet excitons. This phenomenon is expected as a novel process, which can overcome the Shockley/Queisser efficiency limit [3] of organic photovoltaic cell (OPV) due to the following two features.

1. Multiple excitons generation

Multiple excitons generation, in which excitons with a higher energy than twice the band gap separate into two isolated excitons, can produce two electric carriers per one photon. This process benefits the reduction of an energy loss of excitons possessing an excess energy over the band gap (Fig. 5.2) by partly excluding thermalization process. Indeed, Hanna and Nozik have theoretically shown that an OPV composed of optical absorbers, where singlet fission occurs, has the possibility of overcoming the conventional Shockley/Queisser efficiency limit.

2. Long lifetime of triplet exciton

An efficient generation of triplet exciton is another fascinating property of singlet fission. By virtue of the long lifetime of triplet exciton, a diffusion length of triplet exciton is predicted to be quite longer than that of singlet exciton. Najafov and co-workers have experimentally shown that triplet excitons generated by

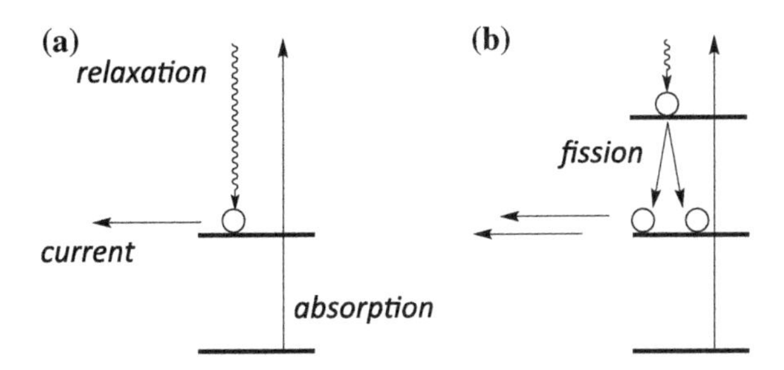

Fig. 5.2 Schematic picture of carrier generation process after absorption of photon for conventional OPV (**a**) and OPV utilizing singlet fission (**b**)

singlet fission in rubrene crystal exhibit an extraordinary long diffusion length (2–8 mm), which is comparable to the optical absorption depth [6]. Therefore, singlet fission is expected to overcome the trade-off problem between exciton diffusion length and optical absorption depth in OPVs, and thus to improve the energy conversion efficiency of photovoltaic system.

Recently, photovoltaic systems utilizing singlet fission have been reported by several authors. Rao and co-workers have shown the first experimental evidence of singlet fission by using the transient absorption spectroscopy for the pentacene/C_{60} photovoltaic cell [7]. Another evidence has been provided by Jadhav and co-workers by using the magnetic response of photocurrent, the characteristic behavior of which determinately indicates the contribution of singlet fission on photoelectric conversion process [8]. In addition to the organic/organic photovoltaic cells, the other types of photovoltaic cells utilizing singlet fission sensitizer (SFS) such as SFS/organic(donor)/organic(acceptor) [9], SFS/quantum dot [10], and SFS/amorphous-silicon [11] have also been proposed. These new types of photovoltaic systems are paving the way for the application of singlet fission to photovoltaic systems.

5.1.2 Kinetic Scheme of Singlet Fission

Singlet fission appears in molecular aggregates and/or crystals, and competes with the other relaxation paths such as internal conversion (IC), intersystem crossing (ISC), fluorescence (Fl), and phosphorescence (Ph) (Fig. 5.3). The kinetic scheme of singlet fission is approximately described by Eq. (5.1.1), where S_n and T_n ($n \geq 1$) indicate the n-th singlet and triplet excited states, respectively, localized on a single molecule, and S_0 denotes the singlet ground state. The intermediate state of singlet fission, $^1(T_1T_1)$, represents a coupled state of two triplet excitons, the overall spin multiplicity of which is singlet (Fig. 5.4). Owing to the spin-allowed process, the transition from $S_0 + S_1$ to $^1(T_1T_1)$ proceeds within several tens ps in tetracene and pentacene, and so on [3]. The time scale of this process is

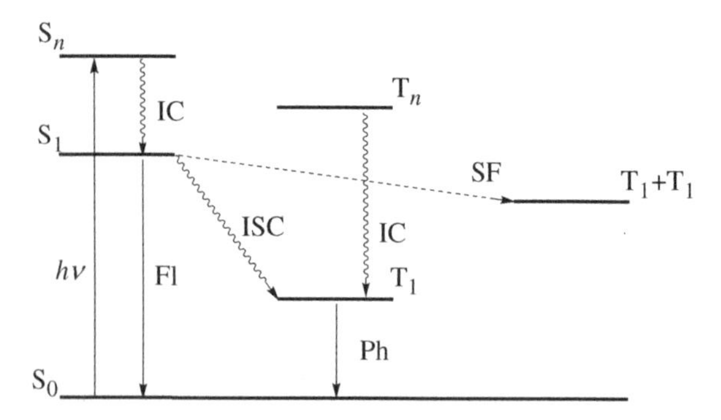

Fig. 5.3 Jablonski diagram with singlet fission (SF), internal conversion (IC), intersystem crossing (ISC), fluorescence (Fl), and phosphorescence (Ph)

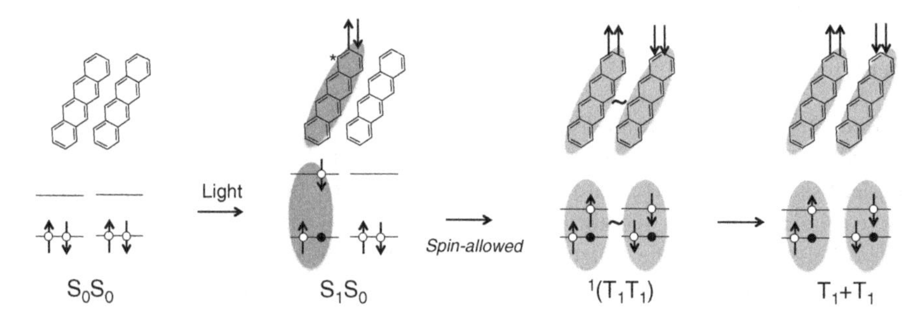

Fig. 5.4 Schematic process of singlet fission. White and black-filled circles denote electron and hole, respectively

considered to be much faster than the other triplet generation process through intersystem crossing (ISC), so that the signal indicating the ultrafast emergence of triplet exciton is often considered as one of the experimental criterion of singlet fission [12]. After the $^1(T_1T_1)$ state is generated, the correlated triplet excitons lose their coherence and separate into non-correlated triplet excitons, denoted as $T_1 + T_1$ in Eq. (5.1.1). Note that, strictly speaking, Eq. (5.1.1) is only an approximate formula for actual process. Indeed, singlet fission sometimes occurs from the higher singlet excited state, S_n ($n > 1$) [13], an exciton might be delocalized over more than two molecules, and/or there might be a mixing of states with different spin multiplicities through a spin-orbit coupling.

Let us consider the wavefunction of a coupled triplet state. For a two-electron system, three triplet eigenfunctions of $\hat{S}^2$ are given by

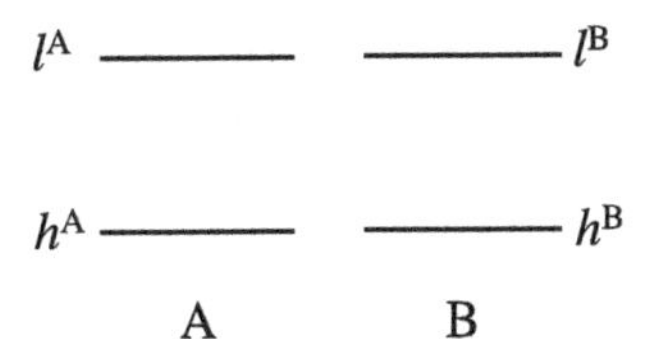

Fig. 5.5 Four electrons in four orbitals in a two-site model

$$
\begin{array}{ll}
|hl\rangle & (S=1,\, M_S=\ +1), \\
\left(|h\bar{l}\rangle + |\bar{h}l\rangle\right)/\sqrt{2} & (S=1,\, M_S=\ 0), \\
|\bar{h}\bar{l}\rangle & (S=1,\, M_S=\ -1),
\end{array}
\tag{5.1.2}
$$

where h and l denote the HOMO (bonding orbital) and LUMO (anti-bonding orbital), respectively. The coupled state is assumed to be composed of two triplet excitons localized on sites A and B, respectively (Fig. 5.5). There are $3 \times 3 = 9$ configurations in total:

$$
\begin{aligned}
& M_S = +2: \\
& \quad \psi_1^{M_S=+2} = \left|h^A l^A h^B l^B\right|. \\
& M_S = +1: \\
& \quad \psi_1^{M_S=+1} = \frac{1}{\sqrt{2}}\left(\left|h^A l^A h^B \bar{l}^B\right| + \left|h^A l^A \bar{h}^B l^B\right|\right), \\
& \quad \psi_2^{M_S=+1} = \frac{1}{\sqrt{2}}\left(\left|h^A \bar{l}^A h^B l^B\right| + \left|\bar{h}^A l^A h^B l^B\right|\right). \\
& M_S = 0: \\
& \quad \psi_1^{M_S=0} = \left|h^A l^A \bar{h}^B \bar{l}^B\right|, \\
& \quad \psi_2^{M_S=0} = \left|\bar{h}^A \bar{l}^A h^B l^B\right|, \\
& \quad \psi_3^{M_S=0} = \frac{1}{2}\left|\left(h^A \bar{l}^A + \bar{h}^A l^A\right)\left(h^B \bar{l}^B + \bar{h}^B l^B\right)\right|. \\
& M_S = -1: \\
& \quad \psi_1^{M_S=-1} = \frac{1}{\sqrt{2}}\left(\left|\bar{h}^A \bar{l}^A h^B \bar{l}^B\right| + \left|\bar{h}^A \bar{l}^A \bar{h}^B l^B\right|\right), \\
& \quad \psi_2^{M_S=-1} = \frac{1}{\sqrt{2}}\left(\left|h^A \bar{l}^A \bar{h}^B \bar{l}^B\right| + \left|\bar{h}^A l^A \bar{h}^B \bar{l}^B\right|\right). \\
& M_S = -2: \\
& \quad \psi_1^{M_S=-2} = \left|\bar{h}^A \bar{l}^A \bar{h}^B \bar{l}^B\right|.
\end{aligned}
\tag{5.1.3}
$$

Some of the configurations in Eq. (5.1.3) are not the eigenvalues of $\hat{S}^2$ operator. By diagonalizing $\hat{S}^2$ matrix, we obtain the following nine eigenvectors both for S^2 and S_z operators:

$$
\begin{aligned}
&{}^5\Psi_1^{M_S=+2} = \psi_1^{M_S=+2}, && (S=2, M_S=+2)\\
&{}^3\Psi_2^{M_S=+1} = \tfrac{1}{\sqrt{2}}\left(\psi_1^{M_S=+1} - \psi_2^{M_S=+1}\right), && (S=1, M_S=+1)\\
&{}^5\Psi_3^{M_S=+1} = \tfrac{1}{\sqrt{2}}\left(\psi_1^{M_S=+1} + \psi_2^{M_S=+1}\right), && (S=2, M_S=+1)\\
&{}^1\Psi_4^{M_S=0} = \tfrac{1}{\sqrt{3}}\left(\psi_1^{M_S=0} + \psi_2^{M_S=0} - \psi_3^{M_S=0}\right), && (S=0, M_S=0)\\
&{}^3\Psi_5^{M_S=0} = \tfrac{1}{\sqrt{2}}\left(\psi_1^{M_S=0} - \psi_2^{M_S=0}\right), && (S=1, M_S=0)\\
&{}^5\Psi_6^{M_S=0} = \tfrac{1}{\sqrt{6}}\left(\psi_1^{M_S=0} + \psi_2^{M_S=0} + 2\psi_3^{M_S=0}\right), && (S=2, M_S=0)\\
&{}^3\Psi_7^{M_S=-1} = \tfrac{1}{\sqrt{2}}\left(\psi_1^{M_S=-1} - \psi_2^{M_S=-1}\right), && (S=1, M_S=-1)\\
&{}^5\Psi_8^{M_S=-1} = \tfrac{1}{\sqrt{2}}\left(\psi_1^{M_S=-1} + \psi_2^{M_S=-1}\right), && (S=2, M_S=-1)\\
&{}^5\Psi_9^{M_S=-2} = \psi_1^{M_S=-2}. && (S=2, M_S=-2)
\end{aligned}
\tag{5.1.4}
$$

These nine wavefunctions are the tetraradical states composed of two triplet wavefunctions, where there are one singlet ($S = 0$), three triplet ($S = 1$), and five quintet ($S = 2$) eigenvectors. The wavefunction of $^1(T_1T_1)$ state, which is the intermediate state of singlet fission, is given by ${}^1\Psi_4^{M_S=0}$ in Eq. (5.1.4).

We next analyze the excited state configurations of the coupled triplet state in MO representation. Let us consider the system with four electrons in four orbitals (Fig. 5.5). First, tetraradical singlet state is considered. We define the semi-localized orbitals (SLOs), ϕ_{XB} and ϕ_{XU} ($X = $ L, R), which localize on the periphery region but delocalize over the two neighboring sites (L, R), by superposing the HOMO and HOMO − 1 and by doing the LUMO and LUMO + 1, respectively (Fig. 5.6):

$$
\phi_{LB} = \frac{1}{\sqrt{2}}(\phi_{HOMO-1} + \phi_{HOMO}), \quad \phi_{RB} = \frac{1}{\sqrt{2}}(\phi_{HOMO-1} - \phi_{HOMO}),
$$

$$
\phi_{RU} = \frac{1}{\sqrt{2}}(\phi_{LUMO} + \phi_{LUMO+1}), \quad \text{and} \quad \phi_{LU} = \frac{1}{\sqrt{2}}(\phi_{LUMO} - \phi_{LUMO+1}) \tag{5.1.5}
$$

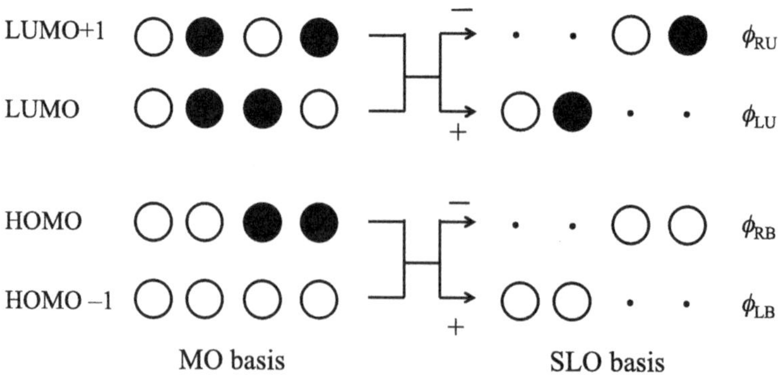

Fig. 5.6 Molecular orbitals (MOs) and semi-localized orbitals (SLOs) basis for four-electron linear model

By using Eq. (5.1.5), we obtain six semi-localized triplet wavefunctions distributed on the left- or right-hand sides:

$$
\begin{aligned}
&{}^{3}\Phi_{\mathrm{L}}^{M_S=+1} = |\phi_{\mathrm{LB}}\phi_{\mathrm{LU}}|,\ {}^{3}\Phi_{\mathrm{L}}^{M_S=-1} = |\bar{\phi}_{\mathrm{LB}}\bar{\phi}_{\mathrm{LU}}|,\ {}^{3}\Phi_{\mathrm{L}}^{M_S=0} = \frac{1}{\sqrt{2}}|\phi_{\mathrm{LB}}\bar{\phi}_{\mathrm{LU}} + \bar{\phi}_{\mathrm{LB}}\phi_{\mathrm{LU}}|,\\
&{}^{3}\Phi_{\mathrm{R}}^{M_S=+1} = |\phi_{\mathrm{RB}}\phi_{\mathrm{RU}}|,\ {}^{3}\Phi_{\mathrm{R}}^{M_S=-1} = |\bar{\phi}_{\mathrm{RB}}\bar{\phi}_{\mathrm{RU}}|,\quad \text{and}\quad {}^{3}\Phi_{\mathrm{R}}^{M_S=0} = \frac{1}{\sqrt{2}}|\phi_{\mathrm{RB}}\bar{\phi}_{\mathrm{RU}} + \bar{\phi}_{\mathrm{RB}}\phi_{\mathrm{RU}}|,
\end{aligned}
\tag{5.1.6}
$$

where the upper-bar represents β spin. From Eq. (5.1.6), we obtain three configurations with $M_S = 0$:

$$
\begin{aligned}
\psi_1^{M_S=0} &= |{}^{3}\Phi_{\mathrm{L}}^{M_S=+1}\,{}^{3}\Phi_{\mathrm{R}}^{M_S=-1}|,\ \psi_2^{M_S=0} = |{}^{3}\Phi_{\mathrm{L}}^{M_S=-1}\,{}^{3}\Phi_{\mathrm{R}}^{M_S=+1}|,\quad \text{and}\\
\psi_3^{M_S=0} &= |{}^{3}\Phi_{\mathrm{L}}^{M_S=0}\,{}^{3}\Phi_{\mathrm{R}}^{M_S=0}|
\end{aligned}
\tag{5.1.7}
$$

By substituting Eqs. (5.1.5) and (5.1.6) into (5.1.7), we obtain

$$
\begin{aligned}
\psi_1^{M_S=0} =& \frac{1}{4}|+13\bar{1}\bar{3} + 13\bar{1}\bar{4} - 13\bar{2}\bar{3} - 13\bar{2}\bar{4} - 14\bar{1}\bar{3} - 14\bar{1}\bar{4} + 14\bar{2}\bar{3} + 14\bar{2}\bar{4}\\
&+23\bar{1}\bar{3} + 23\bar{1}\bar{4} - 23\bar{2}\bar{3} - 23\bar{2}\bar{4} - 24\bar{1}\bar{3} - 24\bar{1}\bar{4} + 24\bar{2}\bar{3} + 24\bar{2}\bar{4}|,
\end{aligned}
\tag{5.1.8}
$$

$$
\begin{aligned}
\psi_2^{M_S=0} =& \frac{1}{4}|+\bar{1}\bar{3}13 + \bar{1}\bar{3}14 - \bar{1}\bar{3}23 - \bar{1}\bar{3}24 - \bar{1}\bar{4}13 - \bar{1}\bar{4}14 + \bar{1}\bar{4}23 + \bar{1}\bar{4}24\\
&+\bar{2}\bar{3}13 + \bar{2}\bar{3}14 - \bar{2}\bar{3}23 - \bar{2}\bar{3}24 - \bar{2}\bar{4}13 - \bar{2}\bar{4}14 + \bar{2}\bar{4}23 + \bar{2}\bar{4}24|,
\end{aligned}
\tag{5.1.9}
$$

$$
\begin{aligned}
\psi_3^{M_S=0} =& \frac{1}{2}|12\bar{3}\bar{4} + 43\bar{2}\bar{1}|\\
&+\frac{1}{4}|-13\bar{1}\bar{3} + 14\bar{1}\bar{4} + 23\bar{2}\bar{3} - 24\bar{2}\bar{4} + 14\bar{2}\bar{3} - 13\bar{2}\bar{4} + 23\bar{1}\bar{4} - 24\bar{1}\bar{3}|,
\end{aligned}
\tag{5.1.10}
$$

where 1, 2, 3, and 4 denote HOMO − 1, HOMO, LUMO, and LUMO + 1, respectively. Equations (5.1.8)–(5.1.10) are not the eigenfunctions of $\hat{S}^2$ operator. The spin adapted wavefunction with singlet spin is obtained by substituting Eqs. (5.1.8)–(5.1.10) into ${}^{1}\Psi_4^{M_S=0}$ in Eq. (5.1.4):

$$
\begin{aligned}
{}^{1}\Psi^{M_S=0} =& \frac{1}{4\sqrt{3}}\{3|13\bar{1}\bar{3} - 14\bar{1}\bar{4} - 23\bar{2}\bar{3} + 24\bar{2}\bar{4}|\\
&-2|12\bar{3}\bar{4} - \bar{1}\bar{2}34| + |\bar{1}2\bar{3}4 + \bar{1}23\bar{4} + 1\bar{2}\bar{3}4 + 1\bar{2}3\bar{4}|\}
\end{aligned}
\tag{5.1.11}
$$

This is the MO representation of a singlet tetraradical state composed of two triplet wavefunctions, where each triplet wavefunction is localized on the left- or right-hand side of the four-site model, the condition of which corresponds to each triplet exciton localized on one molecule. The 1st–4th terms are the perfect-paring type doubly excited configurations, and the 6th–11th terms are the non-perfect paring type doubly excited configurations. It is therefore found that one of the tetraradical singlet states is expressed only by using doubly excited configurations composed of HOMO − 1, HOMO, LUMO, and LUMO + 1. Note that several kinds of tetraradical singlet wavefunctions can be constructed by assuming the different location for two semi-localized triplet wavefunctions.

Similar to the singlet case, a spin-adapted triplet wavefunction with $M_S = +1$ is obtained from ${}^3\Psi_2^{M_S=+1}$ in Eq. (5.1.4):

$$ {}^3\Psi^{M_S=+1} = \frac{1}{2}\{|-1\bar{1}34 + 123\bar{3} - 134\bar{4} + 2\bar{2}34|\}. \quad (5.1.12) $$

Also, spin-adapted quintet wavefunction with $M_S = +2$ is obtained from ${}^5\Psi_1^{M_S=+2}$ in Eq. (5.1.4):

$$ {}^5\Psi^{M_S=+2} = |1234|. \quad (5.1.13) $$

Besides Eqs. (5.1.10)–(5.1.12), the different type of triplet-triplet coupled state can be considered by assuming the different type of SLO. For example, if one superposes the HOMO − 1 and LUMO, and does the HOMO and LUMO + 1, a spin-adapted triplet wavefunction composed of two triplet excitons, $({}^3\Psi^{M_S=+1})'$, is obtained in the same procedure through Eqs. (5.1.5)–(5.1.12):

$$ \left({}^3\Psi^{M_S=+1}\right)' = \frac{1}{2}|-312\bar{2} + 4\bar{4}31 - 41\bar{1}2 + 43\bar{3}2|. \quad (5.1.14) $$

One of the inhibition processes of singlet fission is a fusion of triplet excitons into single exciton (Fig. 5.7). When two triplet excitons approach each other, they form a coupled state with singlet, triplet or quintet multiplicity, expressed as, ${}^1(T_1T_1)$, ${}^3(T_1T_1)$, ${}^5(T_1T_1)$, respectively, whose wavefunctions are given by Eq. (5.1.4). The probabilities to generate singlet, triplet, and quintet states are often predicted to be 1/9, 3/9, and 5/9, respectively, which reflect the numbers of states possessing the corresponding spin multiplicities. These coupled triplet states, ${}^n(T_1T_1)$, subsequently recombine to generate the other single excitons (S_n, T_n, or Q_n) with the corresponding spin multiplicity.

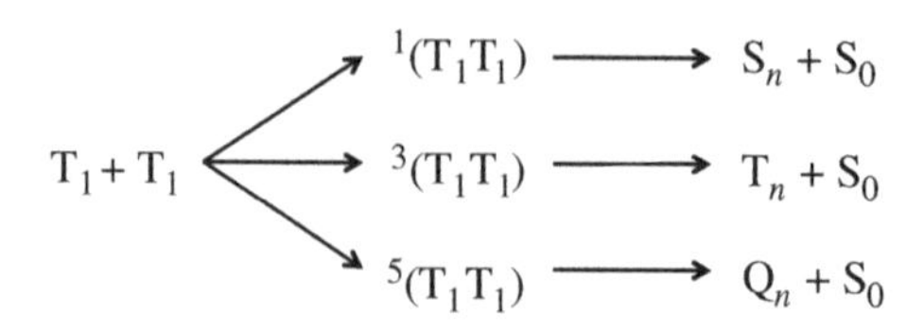

Fig. 5.7 Re-fusion processes of triplet excitons

5.1.3 Molecular Design Guidelines: Energy Level Matching Conditions

A molecular design is one of the important topics in the investigations of singlet fission because only a few numbers of molecules have been known to exhibit singlet fission. To efficiently generate triplet excitons in singlet fission, it is required not only to increase the efficiency of singlet exciton fission but also to suppress the triplet exciton re-fusion. First, in order to increase the efficiency of singlet fission, $S_1 + S_0 \rightarrow T_1 + T_1$ should be an exoergic, an isoergic or at least not a significantly endoergic process, that is, $2E(T_1)–E(S_1) < 0$ (exoergic), $2E(T_1)–E(S_1) \sim 0$ (isoergic or not significantly endoergic), respectively, where $E(S_1)$ and $E(T_1)$ are the excitation energies of the first singlet and triplet excited states, respectively. Although $E(S_1)$ and $E(T_1)$ originally correspond to excitation energies in a molecular crystal or aggregate, we approximately employ the excitation energies of an isolated monomer for simplicity by assuming that an intermolecular interaction is sufficiently weak in molecular crystal. In addition, a highly exoergic process $[2E(T_1) - E(S_1) \ll 0]$ is not preferable for kinetically efficient singlet fission due to the reduction of the reaction rate as shown by the recent discussion on the mechanism of singlet fission [14]. Besides, an excessively exoergic condition is not suitable for the application to OPVs because it leads to a large energy loss in OPVs by wasting a part of energies obtained from an external light. Second, in order to reduce the fusion process of triplet excitons, $T_1 + T_1 \rightarrow S_1 + S_0$, $T_1 + T_1 \rightarrow T_2 + S_0$, and $T_1 + T_1 \rightarrow Q_1 + S_0$ processes should be endoergic, that is, $2E(T_1)-E(S_1) < 0$, $2E(T_1)-E(T_2) < 0$, and $2E(T_1)-E(Q_1) < 0$, respectively. In general, $T_1 + T_1 \rightarrow Q_1 + S_0$ is expected to satisfy an endoergic condition (not proven, but it is empirically assumed), and thus, it is enough to consider only two conditions, $2E(T_1) - E(S_1) < 0$ and $2E(T_1) - E(T_2) < 0$. To summarize the above discussion, the energy level matching conditions for designing a singlet fission molecule are obtained:

$$2E(T_1) - E(S_1) \sim 0, \quad \text{or} \quad 2E(T_1) - E(S_1) < 0 \quad \text{and} \tag{5.1.15}$$

$$2E(T_1) - E(T_2) < 0, \tag{5.1.16}$$

The first condition is required for increasing an efficiency of singlet fission and for suppressing the triplet-triplet annihilation process leading to a singlet exciton (S_1). The second condition is necessary for reducing an efficiency of triplet-triplet re-fusion into the other triplet exciton (T_2). Note that the first condition, Eq. (5.1.15), includes a slightly endoergic condition, as evidenced by the fact that in several molecules such as tetracene singlet fission occurs in endoergic condition. The driving force of singlet fission in such slightly endoergic condition has been recently found to be an entropic gain according to the fission of exciton [15]. Unfortunately, most of organic molecules does not satisfy Eqs. (5.1.15) and (5.1.16) due to their relatively high values of $E(T_1)$. We therefore need a molecular design guideline for reducing $E(T_1)$ as compared to the other excitation energies.

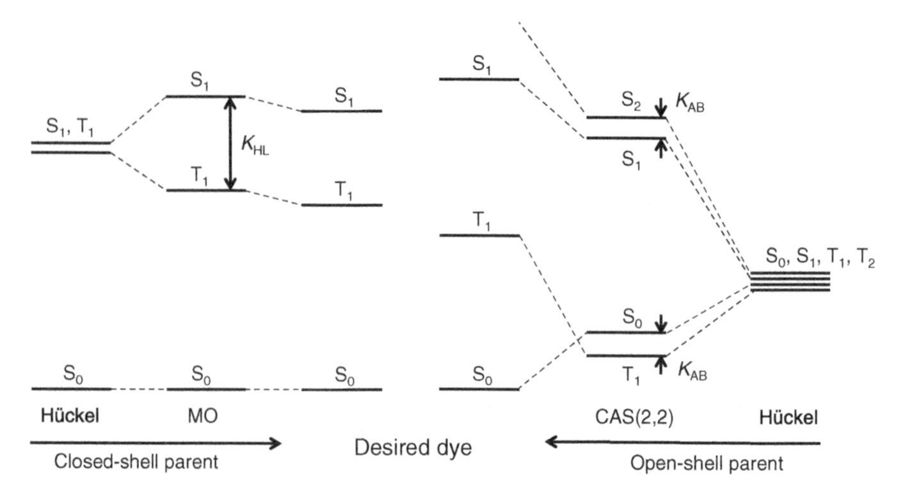

Fig. 5.8 Energy level design of singlet fission sensitizer

In this context, Michl and co-workers have proposed two design guidelines of singlet fission sensitizers by considering the electronic structures of closed-shell and open-shell parent systems (see Fig. 5.8).

5.1.3.1 Alternant Hydrocarbon as a Closed-Shell Parent

For closed-shell parent systems, we assume that the ground state (S_0) is approximately described by a single determinant, and that the S_1 and T_1 states are characterized by the HOMO → LUMO singly excited configuration from the ground state. On the basis of this assumption, we can approximately find that the energy splitting between S_1 and T_1 states is equal to twice the exchange integral between the HOMO and LUMO, which is interpreted as the direct exchange (K_{HL}) between these orbitals, which correlates with their spatial overlap. This indicates that a molecule with large K_{HL} has the possibility of satisfying the energy level matching conditions [Eqs. (5.1.15) and (5.1.16)]. For simplicity, we focus on two typical classes of hydrocarbons: alternant hydrocarbons and non-alternant hydrocarbons. In alternant hydrocarbons, we can apply the starring process: all the carbons in one set are marked with a star, and no two starred or unstarred atoms are bonded with each other (Fig. 5.9a). In the Hückel theory, a corresponding pair of the bonding and antibonding orbitals are known to have mutually the same spatial distribution except for the phase of the wavefunction. This is known as the Coulson–Rushbrooke pairing theorem or alternant pairing theorem [16]. An alternant hydrocarbon will therefore have a large K_{HL} due to the sufficient spatial overlap between the HOMO and LUMO, and will be a possible candidate molecule for satisfying the energy level matching conditions. On the other hand, we cannot apply the starring process to non-alternant hydrocarbons, that is, there exist two neighboring atoms marked by star. In such molecules, the Coulson–Rushbrooke pairing theorem is no longer applicable, and thus, the HOMO and LUMO have

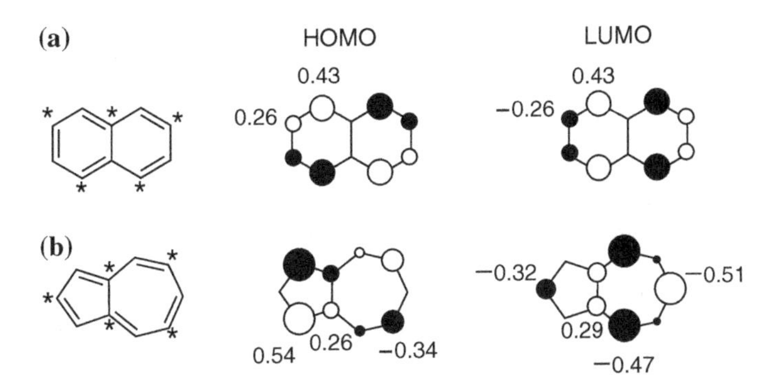

Fig. 5.9 Molecular structures, HOMOs, and LUMOs of alternant hydrocarbons (**a**) and non-alternant hydrocarbon (**b**), calculated using Hückel method

mutually different spatial distributions. As a result, K_{HL} values of non-alternant hydrocarbons are expected to be smaller than those of alternant hydrocarbons. Figure 5.9 shows the HOMO and LUMO of anthracene and azulene, which belong to alternant and non-alternant hydrocarbons, respectively, calculated by using the Hückel method. It is clearly confirmed that the alternant hydrocarbon, anthracene, has the same spatial distributions between the HOMO and LUMO, though the alternant hydrocarbon does not. Therefore, anthracene is predicted to have a larger energy difference between S_1 and T_1 than azulene. By virtue of the pairing orbital distribution, an alternant hydrocarbon is generally predicted to have relatively high K_{HL}, which leads to low T_1 excitation energy as compared to S_1. Of course, more accurate ab initio calculations will give slight different spatial distributions between the HOMO and LUMO even in alternant hydrocarbons, while a qualitative feature of the spatial distributions of these orbitals is expected to be similar to that predicted by the Hückel theory. Indeed, tetracene and pentacene, which belong to alternant hydrocarbons, are known as typical molecules, in which efficient singlet fission occurs.

5.1.3.2 Diradical Molecule as an Open-Shell Parent

In the open-shell parent, where the HOMO and LUMO are degenerate, three singlet (S_0, S_1, S_2) and one triplet (T_1) states have mutually the same energies at the Hückel level of theory. Once an electron repulsion is introduced, the energy levels of S_0, S_1, S_2, and T_1 states are split, and their degeneracy is resolved. The T_1 state approximately lies below the S_0 state by $2K_{AB}$. K_{AB} is the exchange integral between two localized orbitals A and B, each of which is occupied by an unpaired electron. In this pure diradical limit, the T_1 state lies below the S_1 state and satisfy the $2E(T_1) < E(S_1)$, though $E(T_1)$ is too low and is not appropriate for an efficient singlet fission. If one adds a perturbation of changing the molecular structure, that is expected to resolve the degeneracy between the HOMO and LUMO, and to stabilize the S_0 state lower than the T_1 state. Therefore, one can reasonably expect that the

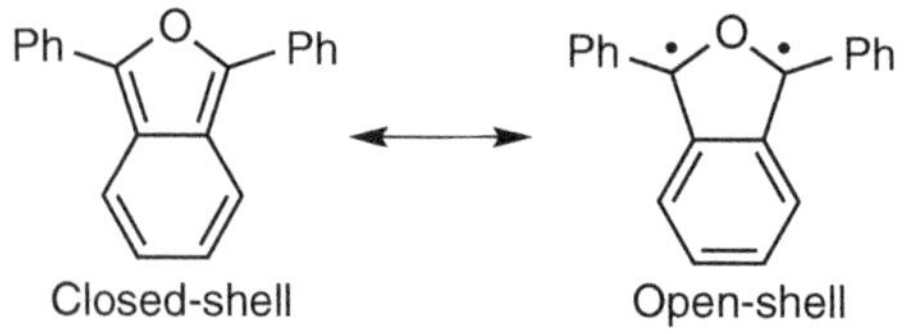

Closed-shell Open-shell

Fig. 5.10 Resonance structures of diphenilisobenzofuran

sufficiently strong perturbation will increase $E(\mathrm{T}_1)$ to satisfy $2E(\mathrm{T}_1) \approx E(\mathrm{S}_1)$. Along this line, Michl et al. have proposed that diphebylisobenzofuran is the promising molecule exhibiting singlet fission, the open-shell character of which is indeed predicted from the open-shell resonance form (Fig. 5.10). However, one should note that not all the diradical molecules are the candidates for efficient singlet fission, because highly diradical molecule is expected to satisfy highly exoergic condition due to its low triplet excitation energy, resulting in significant energy loss and slow transition rate in singlet fission.

Molecules exhibiting singlet fission are classified into three classes of I, II and III, on the basis of an excitation configurations of the first singlet excited state, S_1 (Fig. 5.11) [3]. The appropriate molecular orientation for singlet fission is found to depend on these classes [3]. Class I is characterized by the HOMO $\rightarrow$ LUMO singly excited configuration in S_1 state. Most molecules are classified into this class; for example, tetracene, pentacene, and isobenzofuran belong to class I. On the other hand, class II is characterized by the HOMO–1 $\rightarrow$ LUMO or HOMO $\rightarrow$ LUMO + 1 singly excited configurations, though few molecules are known to belong to this class. From a theoretical prediction, several molecules having a small contribution of zwitterionic resonance structure [17] and polycyclic hydrocarbons with $4n\pi$ electrons [18] are predicted to be classified into this class. In class III, the S_1 state is composed of doubly excited configurations including HOMO $\rightarrow$ LUMO double excitation. Several carotenoids composed of a long polyene unit, such as zeaxanthin, belong to class III. Note that the low-lying singlet doubly excited state is often interpreted as a tetraradical excited state composed of two triplet excitons [19], the wavefunctions of which are given by Eq. (5.1.11).

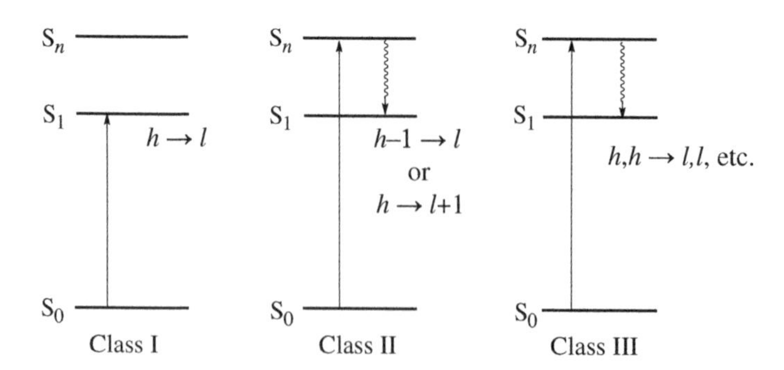

Fig. 5.11 Classes I–III for singlet fission, where h and l indicates the HOMO and LUMO, respectively

5.2 Diradical Character View of Singlet Fission Based on H_4 Model

5.2.1 Introduction

The exploration of singlet fission has been limited so far because singlet fission is observed only in a few numbers of molecules in spite of the pioneering works by Michl et al., who have proposed the strategy to design molecules for singlet fission based on the energy level matching conditions [Eqs. (5.1.15) and (5.1.16)] [3]. Although they have suggested that diradical molecules could be the candidate systems for efficient singlet fission, quantitative discussion on diradical character has never been done. In addition, the effect of the multiple (more than two) singlet open-shell characters such as a tetraradical character on the energy level matching conditions has not been explored yet. The multiple diradical character, which is theoretically defined by the occupation number of the *i*-th lowest unoccupied natural orbital, $y_i = n_{\mathrm{LUNO}+i}(i = 0, 1, \ldots)$, is known to be useful for quantifying the intensities of singlet open-shell nature of multiradical systems [20, 21]. It is also noteworthy that a diradical character correlates with several excitation energies [22, 23]. This suggests that a more sophisticated molecular design guideline for singlet fission can be constructed by introducing a novel viewpoint based on the multiple diradical character. In this section, we aim to establish a novel guideline based on the multiple diradical characters for designing molecules for efficient singlet fission. We investigate the correlation between the multiple diradical characters and excitation energies concerning singlet fission using the symmetric linear H_4 model [24], i.e., a minimum model for tetraradical systems, which are indispensable for investigating singlet fission. The electronic structures are calculated by using the full configuration interaction (full CI) method, which provides an exact solution of the present model. On the basis of the present result, we propose a novel molecular design guideline for singlet fission based on the multiple diradical characters. In addition, the validity of our molecular design guideline is confirmed by evaluating the multiple diradical characters of typical real singlet fission molecules. Finally, several novel candidate molecules for singlet fission are proposed from the viewpoint of the multiple diradical characters.

5.2.2 H_4 Model

We consider a minimal model for singlet fission from the viewpoint of the energy level matching conditions [Eqs. (5.1.15) and (5.1.16)] as well as the classification of singlet fission molecules based on the lowest singlet excited state (S_1). To discuss the energy level matching conditions, we have to clarify the relationship between three excitation energies of $E(S_1)$, $E(T_1)$, and $E(T_2)$. Suppose a T_1 state is described by the HOMO → LUMO singly excited configuration, a T_2 state should

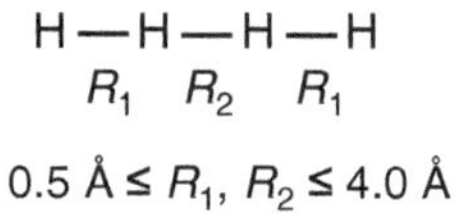

H—H—H—H
R_1 R_2 R_1
0.5 Å ≤ R_1, R_2 ≤ 4.0 Å

Fig. 5.12 Symmetric linear H_4 model

have the other excited configurations concerning HOMO − 1, LUMO + 1, and so on. This implies that at least four electrons are necessary for the description of T_1 and T_2 states. In addition, a minimal model can describe all the classes I, II, and III (see Sect. 5.1.3). Clearly, a four-electron system is enough for describing classes I and II, because it can produce the singly excited configurations of HOMO → LUMO, HOMO − 1→LUMO, and/or HOMO → LUMO + 1, which are included in these classes. On the other hand, the minimum requirement for class III seems to be ambiguous because we do not know how a doubly excited state becomes the lowest singlet excited state. However, if the doubly excited state is considered as a tetraradical excited state, which is composed of two triplet excitons and is described by several doubly excited configurations among HOMO−1, HOMO, LUMO, and LUMO + 1 [Eq. (5.1.11)], it could be the lowest singlet excited state as already demonstrated in polyenes [19]. Therefore, at least four electrons are necessary for the description of class III. To summarize, a four-electron system is reasonable as a minimum model for investigating a molecular design principle for singlet fission. As the simplest example of four-electron system, we consider a symmetric linear H_4 model (Fig. 5.12), in which the singlet open-shell character (diradical and tetraradical characters) can be controlled by changing the interatomic distances, R_1 and R_2 [25].

The multiple diradical characters and excitation energies of the linear H_4 model are calculated using the full CI/STO-3G method with GAMESS program package [26]. The full CI method enables us to obtain the model exact results, which completely include static and dynamic electron correlations. The interatomic distances, R_1 and R_2, are varied from 0.5 to 4.0 Å to span a whole range of multiple diradical characters ($0.0 < y_0, y_1 < 1.0$). For example, the closed-shell, pure diradical, and pure tetraradical ground states [$(y_0, y_1) \sim (0, 0), (1, 0)$, and $(1, 1)$, respectively], are found to be realized by $(R_1, R_2) \sim$ (0.5 Å, 0.5 Å), (4.0 Å, 0.5 Å), and (4.0 Å, 4.0 Å), respectively (see Figs. 5.12 and 5.13).

Before explaining the details of the results, we identify the excited states concerning singlet fission in the H_4 model. The lowest singlet state with A_g symmetry ($S_0 = 1^1A_g$) is the ground state. Similarly, the T_1 and T_2 states are uniquely identified by the 1^3B_u and 1^3A_g states, respectively, where 1^3B_u is the lowest triplet state with B_u symmetry, and 1^3A_g is the lowest triplet state with A_g symmetry. On the other hand, we found that there are two candidates for the first singlet excited state, i.e., $S_1 = 1^1B_u$ or 2^1A_g, where 1^1B_u is the lowest singlet state composed of the HOMO → LUMO singly excited configuration with B_u symmetry, and 2^1A_g is the second lowest singlet state concerning both of the HOMO − 1 (HOMO) → LUMO (LUMO + 1) singly excited configuration and

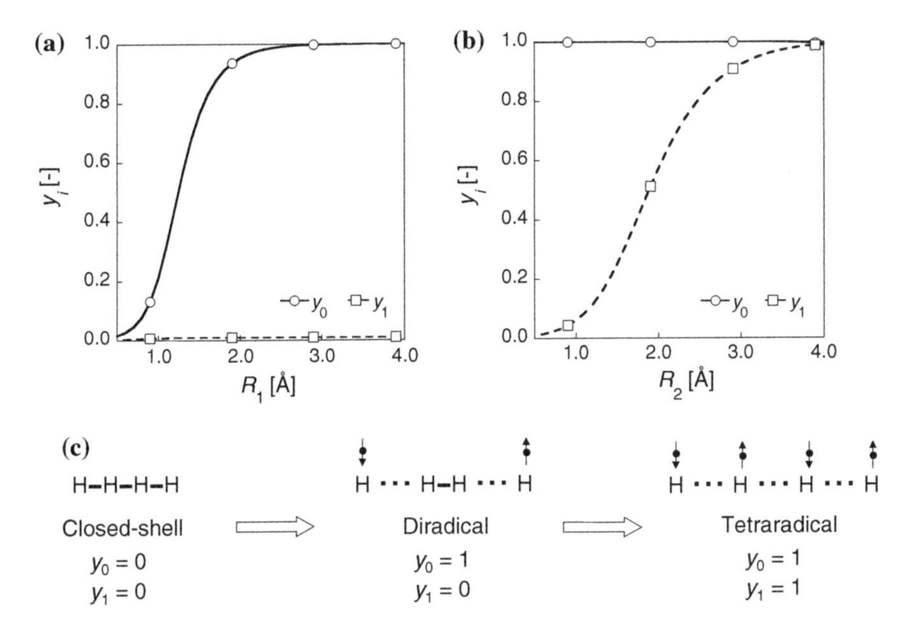

Fig. 5.13 Variations of multiple diradical characters, y_0 and y_1, for the symmetric linear H_4 model as functions of R_1 with fixed $R_2 = 0.5$ Å (**a**), and of R_2 with fixed $R_1 = 4.0$ Å (**b**). A relationship between the H_4 model structure and multiple diradical characters, y_0 and y_1, are also shown (**c**)

several doubly excited configurations with A_g symmetry. According to the classification of molecules for singlet fission (Sect. 5.1.3), the system with $S_1 = 1^1B_u$ belongs to class I, and that with $S_1 = 2^1A_g$ does to class II and/or III in the present model. For simplicity, class III is redefined to include class II, that is, the system with $S_1 = 2^1A_g$ belongs to class III. Note that the order of the 1^1B_u and 2^1A_g states depends on the interatomic distances (R_1 and R_2) as well as on the multiple diradical characters (y_0 and y_1), and thus, both states are considered here. Therefore, the 1^1A_g ground state and the four excited states (1^1B_u, 2^1A_g, 1^3B_u, and 1^3A_g) are essential states for describing the energy level matching conditions as well as for the classification of singlet fission molecules in the present model.

5.2.3 Relationship between Multiple Diradical Character and Excitation Energies in H_4 Model

Figure 5.14 shows the variations in the excitation energies from closed-shell to pure diradical ground states for 0.5 Å $\leq R_1 \leq$ 4.0 Å with fixed $R_2 = 0.5$ Å (a) and from pure diradical to pure tetraradical ground states for 0.5 Å $\leq R_2 \leq$ 4.0 Å with fixed $R_1 = 4.0$ Å (b). Tables 5.1, 5.2, and 5.3 list the excited configurations in the

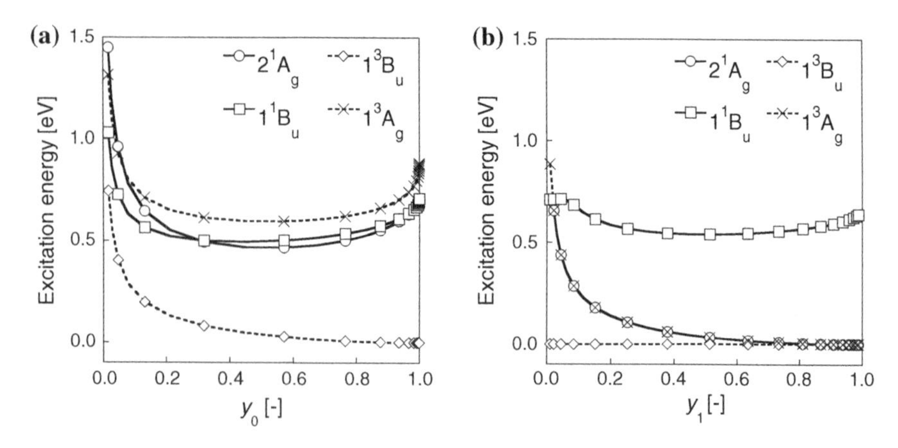

Fig. 5.14 Variations in excitation energies (E) of the 2^1A_g, 1^1B_u, 1^3B_u, and 1^1A_g states as functions of y_0 at $y_1 \sim 0$ (**a**) and of y_1 at $y_0 \sim 1$ (**b**)

Table 5.1 Main configurations at $(R_1, R_2) = (0.5$ Å, 0.5 Å$)$ corresponding to $(y_0, y_1) = (0, 0)$

	CSF[a]	Excitation character[b]		Coefficient
		Occupied	Virtual	
1^1A_g	$\|1\bar{1}2\bar{2}\|$	–	–	0.995
1^1B_u	$\frac{1}{\sqrt{2}}(-\|\bar{3}1\bar{1}2\| + \|31\bar{1}\bar{2}\|)$	h	l	0.994
2^1A_g	$\frac{1}{\sqrt{2}}(-\|\bar{3}12\bar{2}\| + \|3\bar{1}2\bar{2}\|)$	$h-1$	l	0.897
	$\|3\bar{3}1\bar{1}\|$	(h, h)	(l, l)	0.417
1^3B_u	$\|31\bar{1}2\|$	h	l	0.996
1^3A_g	$\|312\bar{2}\|$	$h-1$	l	0.994

[a] Configuration state functional (CSF). (1,2,3,4) denote (HOMO − 1, HOMO, LUMO, LUMO + 1), respectively, and upper bar represents the β spin
[b] h and l denote HOMO and LUMO, respectively

Table 5.2 Main configurations at $(R_1, R_2) = (4.0$ Å, 0.5 Å$)$ corresponding to $(y_0, y_1) = (1, 0)$

	CSF[a]	Excitation character[b]		Coefficient
		Occupied	Virtual	
1^1A_g	$\|1\bar{1}2\bar{2}\|$	–	–	0.705
	$\|3\bar{3}1\bar{1}\|$	(h, h)	(l, l)	−0.705
1^1B_u	$\frac{1}{\sqrt{2}}(-\|\bar{3}1\bar{1}2\| + \|31\bar{1}\bar{2}\|)$	h	l	0.997
2^1A_g	$\|1\bar{1}2\bar{2}\|$	–	–	0.705
	$\|3\bar{3}1\bar{1}\|$	(h, h)	(l, l)	0.705
1^3B_u	$\|31\bar{1}2\|$	h	l	0.997
1^3A_g	$\|312\bar{2}\|$	$h-1$	l	0.999

[a] Configuration state functional (CSF). (1,2,3,4) denote (HOMO − 1, HOMO, LUMO, LUMO + 1), respectively, and upper bar represents the β spin
[b] h and l denote HOMO and LUMO, respectively

Table 5.3 Main configurations at $(R_1, R_2) = (4.0$ Å, 4.0 Å$)$ corresponding to $(y_0, y_1) = (1, 1)$

	CSF[a]	Excitation character[b]		Coefficient
		Occupied	Virtual	
1^1A_g	$\lvert 4\bar{4}2\bar{2}\rvert$	$(h-1, h-1)$	$(l+1, l+1)$	0.249
	$\lvert 1\bar{1}2\bar{2}\rvert$	–	–	−0.504
	$\lvert 1\bar{1}4\bar{4}\rvert$	(h, h)	$(l+1, l+1)$	0.250
	$\frac{1}{2\sqrt{3}}\{-2(\lvert\bar{3}\bar{1}42\rvert + \lvert 31\bar{4}\bar{2}\rvert) + (\lvert\bar{3}1\bar{4}2\rvert + \lvert 3\bar{1}\bar{4}2\rvert + \lvert\bar{3}14\bar{2}\rvert + \lvert 3\bar{1}4\bar{2}\rvert)\}$	$(h-1, h)$	$(l, l+1)$	0.433
	$\frac{1}{2}(\lvert\bar{3}1\bar{4}2\rvert - \lvert 3\bar{1}\bar{4}2\rvert - \lvert\bar{3}14\bar{2}\rvert + \lvert 3\bar{1}4\bar{2}\rvert)$	$(h-1, h)$	$(l, l+1)$	−0.251
	$\lvert 3\bar{3}2\bar{2}\rvert$	$(h-1, h-1)$	(l, l)	0.250
	$\lvert 3\bar{3}4\bar{4}\rvert$	$(h-1, h-1, h, h)$	$(l, l, l+1, l+1)$	−0.495
	$\lvert 3\bar{3}1\bar{1}\rvert$	(h,h)	(l, l)	0.251
1^1B_u	$\frac{1}{\sqrt{2}}(\lvert 1\bar{4}2\bar{2}\rvert - \lvert\bar{1}42\bar{2}\rvert)$	$h-1$	$l+1$	0.312
	$\frac{1}{\sqrt{2}}(\lvert 3\bar{4}2\bar{2}\rvert - \lvert\bar{3}42\bar{2}\rvert)$	$(h-1, h-1)$	$(l, l+1)$	0.253
	$\frac{1}{\sqrt{2}}(\lvert\bar{3}4\bar{4}2\rvert - \lvert 34\bar{4}\bar{2}\rvert)$	$(h-1, h-1, h)$	$(l, l+1, l+1)$	−0.559
	$\frac{1}{\sqrt{2}}(\lvert\bar{3}1\bar{1}2\rvert - \lvert 31\bar{1}\bar{2}\rvert)$	h	l	0.600
	$\frac{1}{\sqrt{2}}(\lvert\bar{3}1\bar{1}4\rvert - \lvert 31\bar{1}\bar{4}\rvert)$	(h, h)	$(l, l+1)$	0.312
	$\frac{1}{\sqrt{2}}(\lvert 3\bar{3}\bar{1}2\rvert - \lvert 3\bar{3}1\bar{2}\rvert)$	$(h-1, h)$	(l, l)	0.255
2^1A_g	$\lvert 4\bar{4}2\bar{2}\rvert$	$(h-1, h-1)$	$(l+1, l+1)$	−0.431
	$\lvert 1\bar{1}4\bar{4}\rvert$	(h, h)	$(l+1, l+1)$	0.433
	$\frac{1}{2\sqrt{3}}\{-2(\lvert\bar{3}\bar{1}42\rvert + \lvert 31\bar{4}\bar{2}\rvert) + (\lvert\bar{3}1\bar{4}2\rvert + \lvert 3\bar{1}\bar{4}2\rvert + \lvert\bar{3}14\bar{2}\rvert + \lvert 3\bar{1}4\bar{2}\rvert)\}$	$(h-1, h)$	$(l, l+1)$	0.250
	$\frac{1}{2}(\lvert\bar{3}1\bar{4}2\rvert - \lvert 3\bar{1}\bar{4}2\rvert - \lvert\bar{3}14\bar{2}\rvert + \lvert 3\bar{1}4\bar{2}\rvert)$	$(h-1, h)$	$(l, l+1)$	0.432
	$\lvert 3\bar{3}2\bar{2}\rvert$	$(h-1, h-1)$	(l, l)	0.433
	$\lvert 3\bar{3}1\bar{1}\rvert$	(h, h)	(l, l)	−0.435
1^3B_u	$\lvert 31\bar{1}2\rvert$	h	l	0.504
	$\lvert 412\bar{2}\rvert$	$h-1$	$l+1$	0.500
	$\lvert 43\bar{3}1\rvert$			0.498

(continued)

Table 5.3 (continued)

	CSF[a]	Excitation character[b]		Coefficient
		Occupied	Virtual	
		$(h-1, h, h)$	$(l, l, l+1)$	
	$\|4\bar{4}32\|$	$(h-1, h-1, h)$	$(l, l+1, l+1)$	0.497
1^3A_g	$\|312\bar{2}\|$	$h-1$	l	0.502
	$\|41\bar{1}2\|$	h	$l+1$	0.502
	$\|43\bar{3}2\|$	$(h-1, h-1, h)$	$(l, l, l+1)$	−0.498
	$\|4\bar{4}31\|$	$(h-1, h, h)$	$(l, l+1, l+1)$	−0.498

[a] Configuration state functional (CSF). (1, 2, 3, 4) denote (HOMO − 1, HOMO, LUMO, LUMO + 1), respectively, and upper bar represents the β spin
[b] h and l denote HOMO and LUMO, respectively

closed-shell [$(R_1, R_2) = (0.5$ Å, 0.5 Å$)$], pure diradical [$(R_1, R_2) = (4.0$ Å, 0.5 Å$)$], and pure tetraradical [$(R_1, R_2) = (4.0$ Å, 4.0 Å$)$] ground states, respectively.

In Fig. 5.14a, $E(1^3B_u)$ shows a characteristic behavior, which decreases with the increase in y_0, and becomes zero at the limit of $y_0 \sim 1$, whereas $E(1^1B_u)$, $E(2^1A_g)$, and $E(1^3A_g)$, are shown to be relatively insensitive to y_0. The decrease behavior of $E(1^3B_u)$ can be understood by the difference in wavefunctions between the 1^1A_g and 1^3B_u states. The wavefunction of 1^1A_g state varies from the closed-shell ($y_0 = 0$) to diradical ($y_1 = 1$), the diradical nature of which is characterized by the doubly excited configuration, $|3\bar{3}1\bar{1}|$, at $(R_1, R_2) = (4.0$ Å, 0.5 Å$)$ (Table 5.2), where (1, 2, 3, 4) denote (HOMO − 1, HOMO, LUMO, LUMO + 1), respectively. On the other hand, the wavefunction of 1^3B_u state, which is an intrinsic diradical state, does not change for $0 < y_0 < 1$ (Tables 5.1 and 5.2). One can therefore expect that the 1^1A_g state has a similar energy to the 1^3B_u state at the limit of $y_0 \sim 1$ because both of them have pure diradical wavefunctions, while the 1^1A_g state has a lower energy than the 1^3B_u state around $y_0 \sim 0$ because the former is stabilized by the bond formation. The present result indicates that the diradical excitation energy, $E(1^3B_u)$, strongly correlates with the index of the diradical nature in the ground state (y_0).

As shown in Fig. 5.14b, $E(2^1A_g)$ and $E(1^3A_g)$ decrease with the increase in y_1, and finally become zero at the limit of $y_1 \sim 1$. These decrease behaviors originate in their tetraradical wavefunctions at $y \sim 1$ because the relative signs and amplitudes of the excited configurations of 2^1A_g and 1^3A_g states at $(R_1, R_2) = (4.0$ Å, 4.0 Å$)$, shown in Table 5.3, are in agreement with those of the analytical solutions of tetraradical states composed of two triplet excitons [Eqs. (5.1.11) and (5.1.14), respectively]. One can therefore expect that the 1^1A_g

state has a similar energy to both the 2^1A_g and 1^3A_g states at $y_1 \sim 1$ because all of them have pure tetraradical wavefunctions, while the ground state (1^1A_g) has a lower energy than the others in the region with small y_1 due to the bond formation. Therefore, the tetraradical excitation energies, $E(2^1A_g)$ and $E(1^3A_g)$, strongly correlate with the index of the tetraradical nature in the ground state (y_1).

In contrast to $E(1^3B_u)$, $E(2^1A_g)$, and $E(1^3A_g)$, $E(1^1B_u)$ is insensitive to the multiple diradical characters (y_0 and y_1), and has non-zero excitation energies in the whole region shown in Fig. 5.14. This behavior originates in the ionic character of the HOMO → LUMO singly excited configuration, which always induces the electron repulsion and destabilized the 1^1B_u state as compared to the ground state (see Tables 5.1, 5.2, and 5.3).

As explained above, it is noteworthy that the multiple diradical character dependences of excitation energies stem from the difference in characters of the wavefunctions between the ground and each excited state.

The comprehensive correlation of $E(1^1B_u)$, $E(2^1A_g)$, $E(1^3B_u)$, and $E(1^3A_g)$ on the multiple diradical characters are shown on the $y_0 - y_1$ plane (Fig. 5.15), where $y_1 \leq y_0$ due to the definition. We clearly found that $E(1^1B_u)$ is insensitive to both y_0 and y_1, and has a non-zero excitation energy in the whole $y_0 - y_1$ region. In contrast, $E(1^3B_u)$ depends on y_0 due to the intrinsic diradical nature of its wavefunction. On the other hand, $E(2^1A_g)$ and $E(1^3A_g)$ depend on y_1 because of the tetraradical nature of their wavefunctions at $(y_0, y_1) = (1, 1)$. The excitation energies of 1^1B_u, 2^1A_g, 1^3B_u, and 1^3A_g states are therefore found to correlate with the multiple diradical characters (y_0 and y_1) through the wavefunctions of excited states similar to the limited case shown in Fig. 5.14.

5.2.4 Diradical Character Based Molecular Design Guideline for Singlet Fission

Figure 5.16 shows the two dimensional maps for the $y_0 - y_1$ plane of the excitation energy differences, $[E(2^1A_g) - E(1^1B_{1u})]$ (a), $[2E(1^3B_{1u}) - E(1^1B_{1u})]$ (b), $[2E(1^3B_{1u}) - E(2^1A_g)]$ (c), and $[2E(1^3B_{1u}) - E(1^3A_g)]$ (d). Figure 5.16a shows $[E(2^1A_g) - E(1^1B_{1u})]$ in order to clarify the class of molecules concerning the S_1 state. The energy difference $[E(2^1A_g) - E(1^1B_u)]$ is found to decrease with the increase in y_1, the tendency of which originates in the correlation between $E(2^1A_g)$ and y_1 as already shown in Fig. 5.15b. Therefore, $S_1 = 2^1A_g$ is realized in the large y_1 region, whereas $S_1 = 1^1B_{1u}$ at $y_1 \sim 0$. This result indicates that the classification of molecules depends on the tetraradical character (y_1).

Figure 5.16b shows the excitation energy difference of $[2E(1^3B_{1u}) - E(1^1B_{1u})]$. The negative value of this quantity implies $2E(T_1) \leq E(S_1)$, which is one of the energy level matching conditions for singlet fission [Eq. (5.1.15)]. This condition is found to be satisfied except for the closed-shell region $[(y_0, y_1) \sim (0, 0)]$ because $[2E(1^3B_{1u}) - E(1^1B_{1u})]$ rapidly decreases and becomes negative in

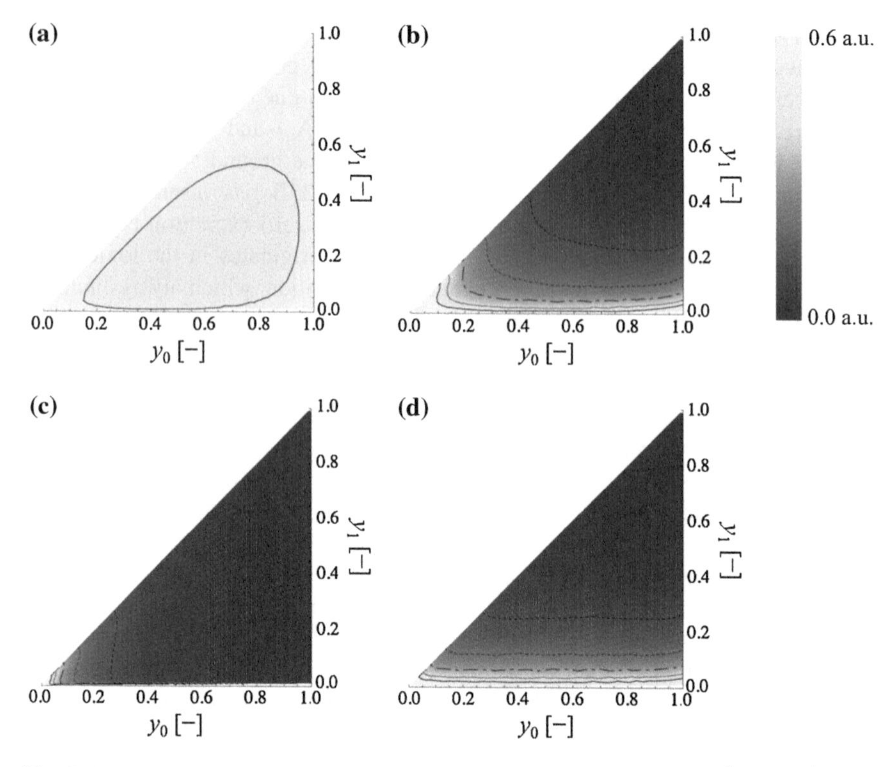

Fig. 5.15 Multiple diradical character dependences of excitation energies of 1^1B_u (**a**), 2^1A_g (**b**), 1^3B_u (**c**), and 1^3A_g (**d**) states in the H_4 model. The *dotted*, *dash-dotted*, and *solid lines* denotes the contour lines of $E = (0.1, 0.2)$, (0.3), and $(0.4, 0.5)$, respectively

sign as increasing y_0, the behavior of which is attributed to the correlation of $E(1^3B_{1u})$ with y_0 as shown in Fig. 5.15c.

Figure 5.16c shows $[2E(1^3B_{1u}) - E(2^1A_g)]$, the negative value of which also represents $2E(T_1) \leq E(S_1)$. Interestingly, $[2E(1^3B_{1u}) - E(2^1A_g)] \sim 0$ is satisfied in a wide range of y_0 for $y_1 > 0.2$ because the 2^1A_g state could be regarded as the tetraradical excited state composed of two triplet excitons in the large y_1 region. On the other hand, we found $[2E(1^3B_{1u}) - E(2^1A_g)] > 0$ around $y_0 \sim 0$ because the 2^1A_g state is no longer the triplet-triplet coupled state as evidenced by the large amount of the ionic contribution of (HOMO − 1 → LUMO) singly excited configuration (Table 5.1).

Figure 5.16d shows $[2E(1^3B_{1u}) - E(2^3A_g)]$. The negative value of this quantity implies $2E(T_1) < E(T_2)$, which is one of the energy level matching conditions for suppressing the re-fusion process of singlet fission [Eq. (5.1.16)]. $[2E(1^3B_u) - E(1^3A_g)]$ is shown to decrease with the increase in y_1 because $E(1^3A_g)$ correlates with y_1 [Fig. 5.15d]. This energy level matching condition is therefore found to be satisfied except for the closed-shell ($y_0 \sim 0$) and tetraradical-like ($y_0 \sim y_1$) regions.

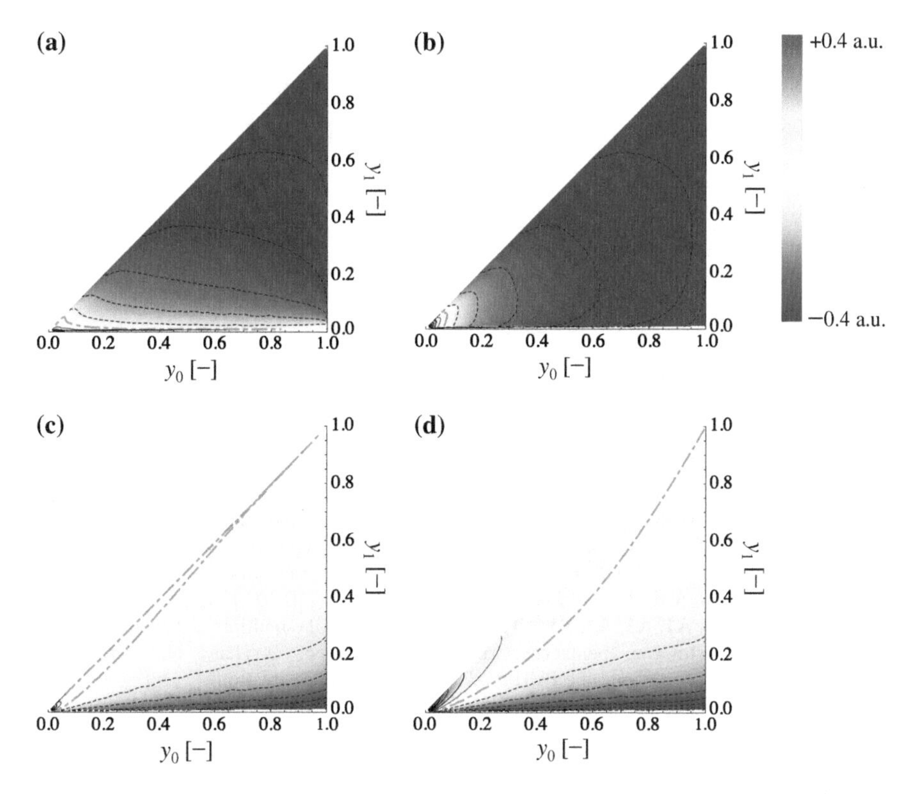

Fig. 5.16 Multiple diradical character dependences of excitation energy differences $[E(2^1A_g) - E(1^1B_{1u})]$ (**a**), $[2E(1^3B_{1u}) - E(1^1B_{1u})]$ (**b**), $[2E(1^3B_{1u}) - E(2^1A_g)]$ (**c**), and $[2E(1^3B_{1u}) - E(1^3A_g)]$ (**d**). The *solid* and *dotted lines* denote the contours with an interval of 0.1 a.u. for $\Delta E > 0$ and $\Delta E < 0$, respectively. The *bold dashed-dotted line* represents the contour with $\Delta E = 0$

These results are summarized in Fig. 5.17, which illustrates the region satisfying the two energy level matching conditions $[2E(T_1) \leq E(S_1)$ and $2E(T_1) < E(T_2)]$ on the y_0-y_1 plane together with the density plot of $h = [2E(1^3B_{1u}) - E(1^1B_{1u})]/E(1^1B_{1u})$, where the darker density exhibits the region with higher energy efficiency ($\sim$0 for the highest efficiency) of singlet fission from the absorption of light by the 1^1B_{1u} state (the lowest optically-allowed state). The auxiliary lines, $[E(2^1A_g) - E(1^1B_{1u})]/E(1^1B_{1u}) = 0$ (solid) and -0.3 (dashed), are also shown in order to clarify the class of molecules for singlet fission.

The energy level matching conditions are found to be unsatisfied by the closed-shell ($y_0 < 0.1$ in the present case) as well as by the tetraradical-like regions ($y_0 \sim y_1$). In addition, the energy efficiency of singlet fission is found to get worse with the increase in y_0 due to the excess decrease of the lowest triplet excitation energy, $E(1^3B_u)$. It is also noteworthy that the doubly excited state (2^1A_g) tends to be the lowest singlet excited state rather than the singly excited state (1^1B_{1u}) in high y_1 region, i.e., class III is predicted to be characterized by the more significant tetraradical nature than class I.

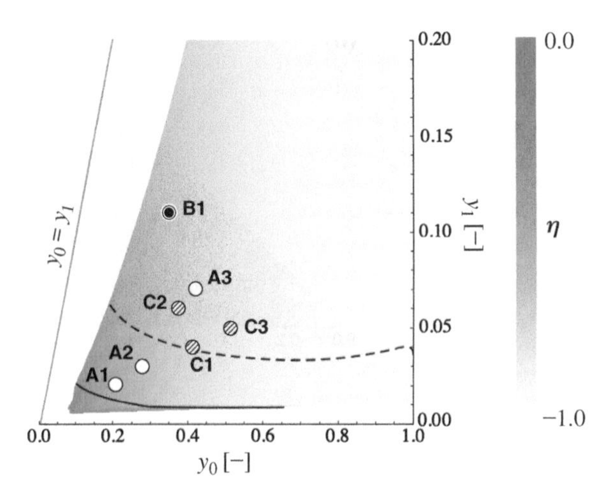

Fig. 5.17 Feasibility region of singlet fission on y_0-y_1 plane on the basis of the energy level matching conditions for the linear H_4 model. The density plot represents the energy efficiency with a function of $h = [2E(1^3B_u) - E(1^1B_u)/E(1^1B_u)]$, where the *darker* region has higher efficiency. The *solid* and *dashed lines* show $[E(2^1A_g) - E(1^1B_u)/E(1^1B_u)] = 0$ and -0.3, respectively. Symbols **A1**, **A2**, **A3**, **B1**, **C1**, **C2**, and **C3** denote the multiple diradical characters of isobenzofuran, tetracene, pentacene, zeaxanthin, zethrene, terrylene, and bisanthene, respectively, calculated by using the PUHF/6-31G* level of approximation

In order to assess the reliability of these results, we examine the multiple diradical characters (Table 5.4) for tetracene (**A1**), 1,3-diphenylisobenzofuran (**A2**), pentacene (**A3**), and zeaxanthin (**B1**) (Fig. 5.18), in which efficient singlet fissions are observed and group “**A**” and “**B**” are found to belong to class I and III, respectively. Their multiple diradical characters are evaluated by using the PUHF/6-31G* method. These molecules are indeed found to lie in the open-shell region with $y_0 > 0.2$. In addition, group **B** exhibits higher y_1 value than group **A**, where the S_1 state of the former is characterized by the doubly excited configurations and that of the latter is done by the HOMO → LUMO singly excited configuration, though group **A** lies slightly outside the region with $E(2^1A_g) - E(1^1B_{1u}) > 0$ in the present model. Nevertheless, the relative features between class I (**A1**, **A2, A3**) and III (**B1**) are distinguished on the y_0-y_1 plane. These results indicate that the multiple diradical characters are useful for the semi-quantitative prediction of the feasibility of singlet fission, while it is still difficult to predict the precise thresholds of multiple diradical characters due to the limitation of the minimum H_4 model and/or to the approximate calculation of the multiple diradical characters. Finally, it is noteworthy that the three polycyclic aromatic hydrocarbons, i.e., zethrene (**C1**), terrylene (**C2**) and bisanthene (**C3**) lie in the feasibility region of singlet fission (see Table 5.4 and Fig. 5.17) though they have not been examined from the viewpoint of singlet fission yet. As seen from Fig. 5.17, these PAHs are predicted to lie in the boundary region between the class I and III due to their intermediate y_0 and non-negligible y_1 values. Also, judging from the increase

Table 5.4 Multiple diradical characters (y_0, y_1)[a] and primary excited configurations in the lowest singlet excited states (S_1) of several real π-conjugated molecules

Molecule	y_0	y_1	Primary configuration of S_1 state
Tetracene[b]	0.28	0.03	HOMO → LUMO single excitation[d]
1,3-Diphenyliso-benzofran[b]	0.21	0.02	HOMO → LUMO single excitation[e]
Pentacene[b]	0.42	0.07	HOMO → LUMO single excitation[f]
Zeaxanthin[b]	0.35	0.11	Double excitation[g]
Zethrene[b]	0.41	0.04	–
Terrylene[c]	0.37	0.06	–
Bisanthene[c]	0.51	0.05	–

[a] Obtained by the PUHF/6-31G* calculations [Eq. (2.2.36)]
[b] Geometries are optimized by the RB3LYP/6-311G* level of theory
[c] Ref. [27]
[d] Ref. [28]
[e] Ref. [29]
[f] Ref. [3]
[g] Ref. [30]

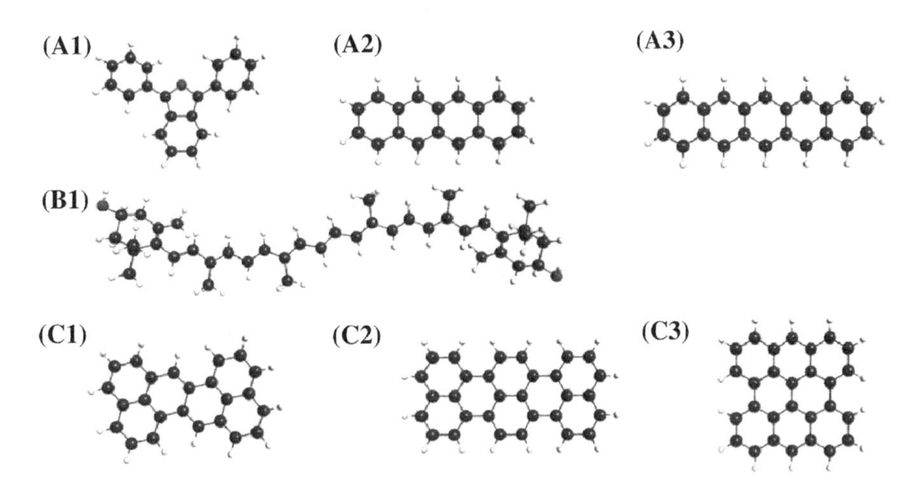

Fig. 5.18 Molecules in which efficient singlet fissions are observed (class I: **A1**, **A2**, and **A3**, class III: **B1**), and novel candidate molecules for singlet fission on the basis of the multiple diradical character based design principle (**C1**, **C2**, and **C3**)

behavior of y_0 and y_1 depending on the size of PAHs, several large-size PAHs are expected to exhibit efficient singlet fission as class III molecules. To verify this prediction, we need to perform strongly electron-correlated calculations of the low-lying excited states in large size PAHs using the multi-reference perturbation

(MRMP), coupled cluster (MRCC), and spin-flip time-dependent DFT methods, which are also challenging themes in quantum chemistry.

In summary, we present a novel viewpoint of multiple diradical character for the energy level matching conditions of singlet fission. It is found that the multiple diradical character is an effective indicator for exploring molecular systems for singlet fission. On the basis of the multiple diradical character, we propose the following guidelines for designing molecules for efficient singlet fission:

- Energy level matching conditions are satisfied by the systems except for closed-shell ($y_0 \sim 0$ and $y_1 \sim 0$) and tetraradical-like ($y_0 \sim y_1$) systems.
- Energy loss in the process of singlet fission increases with the increase in y_0.
- Weak/intermediate diradical molecules are possible candidates for singlet fission with small energy loss.
- Tetraradical character (y_1) affects the classification of singlet fission molecules, for example, highly tetraradical molecules tend to belong to class III.

It is especially noteworthy that efficient singlet fission is predicted to occur in molecules with weak/intermediate diradical characters rather than those with high singlet open-shell or closed-shell characters. Such molecular systems with small/intermediate open-shell singlet characters are recently predicted to be realized in several PAHs [31, 32] and in other thermally stable compounds including diphenalenyl diradicaloids [33, 34]. Among them, we have predicted that the relatively small-size oligorylenes [35–39] with weak/intermediate diradical characters, i.e., terrylene and quaterrylene, are suitable for energetically efficient singlet fission based on the energy level matching conditions [40] using the TD-tuned-LC-BLYP [41, 42] and PUHF methods. The present results will therefore stimulate theoretical and experimental investigations on singlet fission in open-shell singlet molecules from the viewpoint of multiple diradical characters, which pave the way towards realizing highly efficient OPVs.

5.3 [n]Acenes and [n]Phenacenes

5.3.1 Introduction

The first molecular design guidelines for singlet fission have been proposed by Michl et al., who predicted that alternant hydrocarbons (i) or biradicaloids (ii) are the possible candidates for singlet fission [3]. Suppose both the first singlet (S_1) and triplet (T_1) excited states are described by the HOMO → LUMO singly excited configuration, the energy difference between these states is approximately in proportion to K_{HL} [43], which is the exchange integral between the HOMO and LUMO. In alternant hydrocarbons, the K_{HL} is generally expected to be larger than

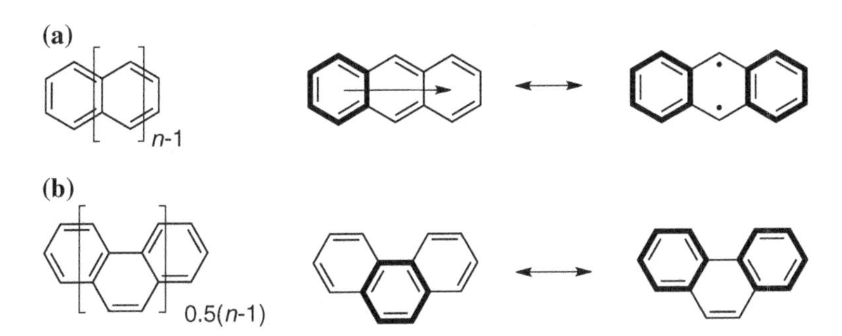

Fig. 5.19 Molecular structures of polyacenes (**a**) and polyphenacenes (**b**) as well as the resonance structures of anthracene and phenacenes as examples of **a** and **b**, respectively

non-alternant hydrocarbons, owing to the large spatial overlap between the HOMO and LUMO as predicted by the Coulson–Rushbrooke pairing theorem in the Hückel theory [16]. Therefore, one can expect that alternant hydrocarbons have the possibility of satisfying one of the energy level matching conditions, $2E(T_1) \sim E(S_1)$ or $2E(T_1) < E(S_1)$ [Eq. (5.1.15)], which is required for realizing singlet fission under isoenergetic or exoenergetic condition, respectively. Indeed, tetracene and pentacene, which belong to alternant hydrocarbons, are known to exhibit efficient singlet fission [44]. However, there are some exceptions such as [n]phenacenes, which do not exhibit efficient singlet fission though they belong to alternant hydrocarbons and have the same molecular formulas as the corresponding [n]acenes (Fig. 5.19).

In this section, the diradical character based design principle (Sect. 5.2) is applied to two kinds of typical alternant hydrocarbons, [n]acenes and [n]phenacenes, in order to clarify the reason why some of the formers exhibit singlet fission though the latters do not. In addition, we also illuminate how a diradical character works on one of the energy level matching conditions, $2E(T_1) \sim E(S_1)$ or $2E(T_1) < E(S_1)$, in order to provide the theoretical rationality and the reliability of diradical character based design principle. Note here that the first condition will be useful for the designing singlet fission molecules since relatively small-size hydrocarbon systems with small/intermediate y_0 values, which satisfy the first condition, mostly tend to give much smaller y_1 values ($y_1 << y_0$), which satisfy the second condition [40]. Thus, we discuss the analytical expressions of diradical character and excitation energy difference [$2E(T_1) - E(S_1)$] concerning the energy level matching condition in the two-site diradical model. Then, the origin of the difference between [n]acenes and [n]phenacenes is unveiled from the viewpoint of the diradical characters [45].

5.3.2 Analytical Relationship between Energy Matching Condition and Diradical Character

First, the analytical expressions of the excitation energies and diradical character by the VCI method [22, 23] are briefly explained as follows. As the simplest diradical system, we consider a two-site system, $A^{\bullet} - B^{\bullet}$, with two electrons in two active orbitals. Using the spatial symmetry-adapted bonding (g) and anti-bonding (u) molecular orbitals (MOs) obtained by the highest spin (triplet in the present case) state spin-unrestricted (U) Hartree-Fock (HF) or density functional theory (DFT) method, the localized natural orbitals (LNOs) can be defined by Eq. (3.1.2). In the VCI method, the diradical character y is given in the LNO basis representation by [22]

$$y = 2C_{\mathrm{D}}^2 = 1 - \frac{1}{\sqrt{1 + \left(\frac{U}{4|t_{ab}|}\right)^2}}, \tag{5.3.1}$$

where C_{D} is the coefficient of a doubly excited configuration in the lowest singlet state $(= C_{\mathrm{G}}|g\bar{g}\rangle + C_{\mathrm{D}}|u\bar{u}\rangle)$. As increasing the weight of doubly excited configuration, a chemical bond is gradually weakened, and is finally broken in the case of $|C_{\mathrm{G}}| = |C_{\mathrm{D}}| = 1/\sqrt{2}$. Therefore, $y = 2C_{\mathrm{D}}^2 = 0$ and 1 indicate closed-shell and pure diradical ground states, respectively. Note that $|t_{ab}|$ and U $(=U_{aa} - U_{ab})$ in Eq. (5.3.1) is defined by the LNO basis (a and b). By using Eq. (3.1.2), $|t_{ab}|$ and U are converted into the MO basis (g and u):

$$2t_{ab} = -(\varepsilon_g^{\mathrm{LNO}} - \varepsilon_u^{\mathrm{LNO}}) = -\Delta\varepsilon_{gu}^{\mathrm{LNO}}, \tag{5.3.2}$$

$$U = U_{aa} - U_{ab} = (aa|aa) - (aa|bb) - 2(gu|ug) = 2K_{gu}. \tag{5.3.3}$$

The derivation of Eq. (5.3.2) is shown in ref. [46]. Note that $\Delta\varepsilon_{gu}^{\mathrm{LNO}}$ is the g-u orbital energy gap defined by the Fock matrix in the LNO basis (see Sect. 2.1), and thus is not identical with the HOMO–LUMO gap of singlet state [46]. Indeed, at the HF level, $2t_{ab}$ is obtained by the orbital energy gap ($\Delta\bar{\varepsilon}_{gu}^{\mathrm{UHF(T)}}$) averaged over α and β orbital gaps in the triplet UHF solution [46]. However, $\Delta\bar{\varepsilon}_{gu}^{\mathrm{UHF(T)}}$ practically shows a linear dependence on the HOMO–LUMO gap of singlet state [46]. Keep in mind that the right-hand side of Eqs. (5.3.2) and (5.3.3) are the expressions in the MO basis representation, which are different from those in the LNO basis representation; for example, K_{ab} and K_{gu} are mutually different because the former is represented by the LNO basis but the latter is done by the MO basis. By substituting Eqs. (5.3.2) and (5.3.3) into Eq. (3.1.2), y is converted from the LNO to MO bases:

$$y = 1 - \frac{1}{\sqrt{1 + \left(\frac{K_{gu}}{\Delta\varepsilon_{gu}^{\mathrm{LNO}}}\right)^2}}. \tag{5.3.4}$$

Equation (5.3.4) explicitly shows that diradical character y is a function of exchange integral (K_{gu}) and g-u orbital energy gap ($\Delta\varepsilon_{gu}^{\mathrm{LNO}}$) of the Fock matrix in the LNO representation, that is, y increases with the increase in K_{gu} and/or the decrease in $\Delta\varepsilon_{gu}^{\mathrm{LNO}}$. This indicates that the concept of one of the conventional molecular design guidelines on alternant hydrocarbons, which are expected to have large K_{HL} (=K_{gu}, in this case), is found to be described by the increase in y with keeping $\Delta\varepsilon_{gu}^{\mathrm{LNO}}$ almost constant.

The g-u single excitation energies of singlet (E_{S}) and triplet (E_{T}) states in the two-site diradical model are given by [46]

$$E_{\mathrm{S}} = \frac{U}{2}\left(1 + \frac{1}{\sqrt{1-(1-y)^2}}\right) - 2K_{ab}, \quad \text{and}$$
$$E_{\mathrm{T}} = \frac{U}{2}\left(-1 + \frac{1}{\sqrt{1-(1-y)^2}}\right) - 2K_{ab}, \tag{5.3.5}$$

The non-dimensional evaluating function of energy level matching condition [$2E(\mathrm{T}_1) \sim 2E(\mathrm{S}_1)$ or $2E(\mathrm{T}_1) < 2E(\mathrm{S}_1)$] in two-site diradical model can be obtained by assuming $E_{\mathrm{S}} = E(\mathrm{S}_1)$ and $E_{\mathrm{T}} = E(\mathrm{T}_1)$, and by substituting Eq. (5.3.5) into $f = [2E(\mathrm{T}_1) - E(\mathrm{S}_1)]/U$,

$$f = -\frac{3}{2} + \frac{1}{2\sqrt{1-(1-y)^2}} - \frac{2K_{ab}}{U}. \quad (f \geq -1) \tag{5.3.6}$$

where $f \geq -1$ is obtained by assuming a singlet ground state [$E(\mathrm{T}_1) \geq 0$] in Eq. (5.3.5). As seen from Eq. (5.3.6), f decreases with the increase in y, and thus, a molecule with a large y value has the possibility to satisfy $2E(\mathrm{T}_1) \sim 2E(\mathrm{S}_1)$ or $2E(\mathrm{T}_1) < 2E(\mathrm{S}_1)$. The third terms of Eq. (5.3.6) ($2K_{ab}/U$) are predicted to be small in most diradical molecules because K_{ab} [=$(ab|ab)$] depends on the spatial overlap between the unpaired electrons though U [=$(aa|aa) - (aa|bb)$] does on the Coulomb interaction between them, the former and latter of which are short- and long-range interaction, respectively. In particular, this prediction, that is, $2K_{ab}/U$ is small, seems to be satisfied for alternant hydrocarbons because K_{ab} is expected to be quite small due to the pairing character in the HOMO and LUMO, which leads to the negligible overlap between the corresponding LNOs [16, 17]. In fact, U and K_{ab} of anthracene are estimated to be 2.07 and 0.03 eV, respectively, at the CASCI/6-31G* level of theory with two-active electrons in two active orbitals by using Eqs. (2.1.9) and (2.1.10). In such cases, y is found to be the controlling factor of one of the energy level matching conditions, $2E(\mathrm{T}_1) \sim 2E(\mathrm{S}_1)$ or $2E(\mathrm{T}_1) < 2E(\mathrm{S}_1)$, when both the S_1 and T_1 states are described by the HOMO → LUMO singly excited configuration.

By assuming one of the energy level matching conditions ($f \sim 0$ or $f < 0$), we obtain

$$E(\mathrm{S}_1) \sim 2U \quad \text{and} \quad E(\mathrm{T}_1) \sim U \quad \text{or} \quad E(\mathrm{S}_1) < 2U \quad \text{and} \quad E(\mathrm{T}_1) < U. \tag{5.3.7}$$

Although this is the natural consequence derived from the functional forms of U [$=E(\mathrm{S}_1) - E(\mathrm{T}_1)$] and the energy level matching condition [$=2E(\mathrm{T}_1) - E(\mathrm{S}_1) \sim 0$ or < 0], Eq. (5.3.7) provides the important understanding that U determines the maximum limit of $E(\mathrm{S}_1)$ and $E(\mathrm{T}_1)$ for a singlet fission molecule. A sufficiently large $E(\mathrm{T}_1)$ would be favorable to a singlet fission photovoltaic system, because $E(\mathrm{T}_1)$ determines the driving force of the charge generation. According to the theoretical prediction for an ideal condition, the appropriate $E(\mathrm{T}_1)$ is indeed predicted to be about 1 eV [3, 5]. In such a system, the third term of Eq. (5.3.6) ($2K_{ab}/U$) should be small due to the relatively large U value. As a result, y_0 is expected to be a particularly important factor for tuning f [Eq (5.3.6)] to satisfy the energy level matching condition, $f \sim 0$ or $f < 0$.

5.3.3 Computational Details

Geometries of the series of [*n*]acenes and [*n*]phenacenes are optimized by using the spin-flip (SF) time-dependent density functional theory (TDDFT) method with the Tamm-Dancoff approximations [47] (TDA) as well as with the collinear approximation [48]. The BHHLYP functional (50 % the HF plus 50 % the Becke exchange [49] with the Lee-Yang-Parr correlation [50]) and the 6-311G* basis sets are employed in this calculation. The reference states for the spin-flip method (T_1 with $M_S = 1$ spin multiplicity) are obtained at the restricted open-shell level of theories. The spatial symmetries of [*n*]acenes and [*n*]phenacenes are fixed to be D_{2h} and C_{2v}, respectively.

The excitation energies of [*n*]acenes and [*n*]phenacenes are calculated using the spin-flip TDDFT/TDA method with the noncollinear approximation [51]. The long-range corrected (LC)-BLYP/6-31G* method with the range-separating parameter of $\mu = 0.33$ bohr^{-1} [52] is employed in order to avoid the error in a global hybrid DFT for the HOMO → LUMO single excitation energies of a series of acenes [53]. The reference states (T_1 with $M_S = 1$) are calculated using the spin-unrestricted approach. For [*n*]acenes and [*n*]phenacenes, the reference states are calculated so as to agree with $\mathrm{b}_{2g} \rightarrow \mathrm{b}_{3u}$ and $\mathrm{b}_1 \rightarrow \mathrm{a}_2$ excited configurations from the singlet ground states, respectively. In this study, the HOMO → LUMO single excitation energies of singlet ($^1E_{\mathrm{HL}}$) and triplet ($^1E_{\mathrm{HL}}$) states are only focused for the comparison with the two-site diradical model. Although

[n]phenacenes have the lowest singlet excited state with other excited configurations, this is nearly degenerate with the HOMO → LUMO singlet excited state, and thus, the present result is not affected.

Effective Coulomb repulsions (U) and transfer integrals (t_{ab}) of [n]acenes and [n]phenacenes are approximately estimated by using the UDFT and SFTDDFT methods combined with the following relationship [46]:

$$U = {}^{1}E_{\mathrm{HL}} - {}^{3}E_{\mathrm{HL}}, \tag{5.3.8}$$

and

$$t_{ab} = -\frac{1}{2}\Delta\bar{\varepsilon}_{\mathrm{HL}}^{\mathrm{T}}. \tag{5.3.9}$$

Diradical characters (y) of [n]acenes and [n]phenacenes are calculated by using the estimated U and t_{ab}. Geometry optimizations and excited state calculations were performed by using the GAMESS program package [26] and Q-CHEM 4.0 program package [54], respectively.

5.3.4 Difference in Excitation Energies between [n]Acenes and [n]Phenacenes

Figure 5.20a shows the HOMO → LUMO single excitation energies of singlet (${}^{1}E_{\mathrm{HL}}$) and triplet (${}^{3}E_{\mathrm{HL}}$) states for [n]acenes and [n]phenacenes. The discrepancy between [n]acenes and [n]phenacenes is explicitly found, that is, ${}^{1}E_{\mathrm{HL}}$ and ${}^{3}E_{\mathrm{HL}}$ of

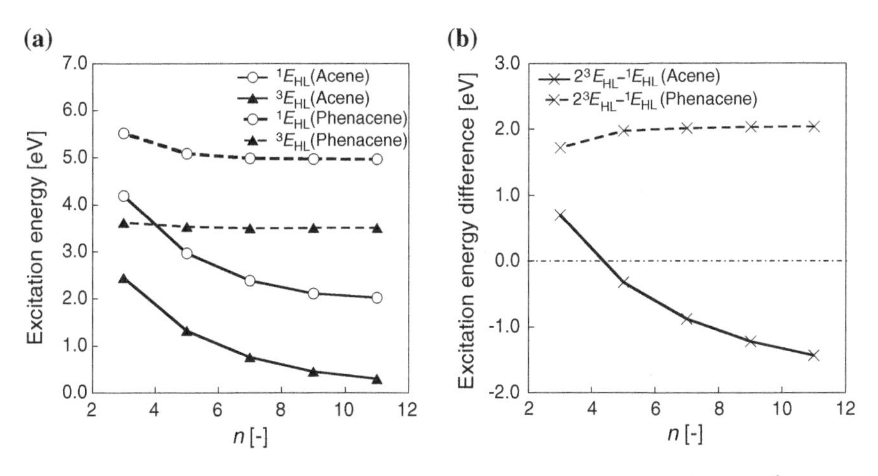

Fig. 5.20 Variations in the HOMO → LUMO single excitation energies of singlet (${}^{1}E_{\mathrm{HL}}$) and triplet states (${}^{3}E_{\mathrm{HL}}$) (**a**), and their differences ($2\,{}^{3}E_{\mathrm{HL}} - {}^{1}E_{\mathrm{HL}}$) concerning the energy level matching condition [Eq. (5.1.15)] (**b**) as the functions of molecular size n for [n]acenes and [n]phenacenes

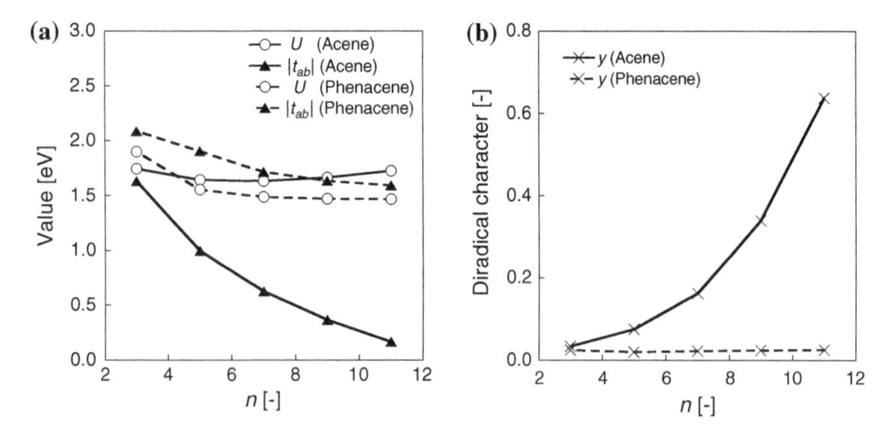

Fig. 5.21 Variations in U and $|t_{ab}|$ estimated from Eqs. (5.3.2) and (5.3.3), respectively (**a**), and diradical character y (**b**) as the functions of molecular size n for [n]acenes and [n]phenacenes

the formers decrease with the increase in the molecular size (n), while those of the latters do not. As a result, the excitation energy differences between $^1E_{\mathrm{HL}}$ and $^3E_{\mathrm{HL}}$, which concern the energy level matching condition, $2E(\mathrm{T}_1)-E(\mathrm{S}_1)$, are also illuminated between [n]acenes and [n]phenacenes. Importantly, [n]acenes satisfy the energy level matching condition, $2E(\mathrm{T}_1)-E(\mathrm{S}_1) \sim 0$ or $2E(\mathrm{T}_1)-E(\mathrm{S}_1) < 0$, for $n > 5$ in the present calculation, while [n]phenacenes do not in the whole size region. This result is in quantitative agreement with the fact that some of acenes are experimentally observed to show singlet fissions though there are few reports of singlet fission in a series of phenacenes.

To clarify the origin of the discrepancy of excitation energies between [n]acenes and [n]phenacenes, Fig. 5.21a shows the U and $|t_{ab}|$ as the function of the molecular size (n). Both [n]acenes and [n]phenacenes show slight dependences of U on the molecular size because U ($=U_{aa} - U_{ab}$) depends on the Coulomb interaction between LNOs a and b (localized in the latitudinal direction), which is the long-range interaction and thereby is predicted to be insensitive to the molecular size (in the longitudinal direction). The mutually similar U values of [n]acenes and [n]phenacenes reflect the fact that both molecules belong to alternant hydrocarbons, which are expected to have large K_{HL} [$=U/2$, see Eq. (5.3.3)] due to the pairing character of HOMO and LUMO predicted by the Coulson–Rushbrooke pairing theorem. [n]Acenes show slight increases in U for $n > 7$ probably due to the excess localization of orbitals induced by the emergence of singlet open-shell character and/or spin contamination. More importantly, the considerable difference between [n]acenes and [n]phenacenes is found to appear in $|t_{ab}|$: $|t_{ab}|$ of [n]acenes significantly decreases with the increase in the molecular size (n) though that of [n]phenacenes do not so much. This discrepancy in $|t_{ab}|$ causes the difference in the diradical character y (Fig. 5.21b), in which the y values of [n]acenes increase with the increase in the molecular size (n), while those of

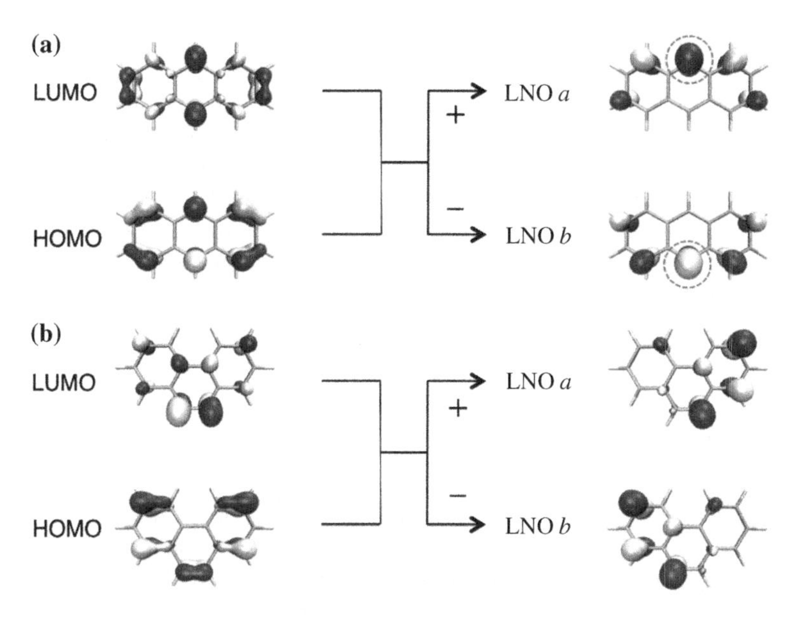

Fig. 5.22 HOMO and LUMO with α spin, and LNOs α and β of anthracene **a** and phenanthrene **b** calculated by using the LC-UBLYP/6-31G* method. The contour values are 0.06 a.u. The atoms indicated by *dotted circles* do not have significant orbital overlaps with the neighboring atoms of the counterpart LNO

[*n*]phenacenes do not. As shown in Eq. (5.3.7), the increase in y is expected to decrease $2^3E_{\mathrm{HL}} - {}^1E_{\mathrm{HL}}$ and to affect the energy level matching condition. Therefore, the difference in $2^3E_{\mathrm{HL}} - {}^1E_{\mathrm{HL}}$ between [*n*]acenes and [*n*]phenacenes is found to originate from $|t_{ab}|$, which is the transfer integral between LNOs a and b, and also correlates with the HOMO–LUMO gap [35].

Figure 5.22 shows the HOMO, LUMO, and the corresponding LNOs a and b of anthracene (a) and phenanthrene (b) in order to clarify the origin of the difference in $|t_{ab}|$ between [*n*]acenes and [*n*]phenacenes. The dotted circle represents the atom with no significant orbital overlap with the neighboring atom of counterpart LNO, and contributes to the reduction of $|t_{ab}|$. Note that the dotted circle only appears for anthracene, but not for phenanthrene. In anthracene (Fig. 5.22a), we found that most of the orbital overlap between LNOs a and b originate in the atoms located in the terminal regions as evidenced by the disappearance of dotted circles in these regions. This result predicts that the longer [*n*]acenes give the smaller $|t_{ab}|$ due to the decrease in the orbital distribution of the LNOs on the terminal atoms. This prediction is in agreement with the computational results shown in Fig. 5.22a, in which $|t_{ab}|$ of [*n*]acenes significantly decreases with the increase in the molecular size (n). On the other hand, for phenanthrene (Fig. 5.22b), almost all the orbitals are found to contribute to the increase in $|t_{ab}|$ due to the no regions indicated by dotted circles. This indicates that the molecular size dependence of $|t_{ab}|$ for [*n*]phenacenes originates only in the delocalization of orbitals and thereby is not significant compared to that of [n]acenes, in which spatially separated unpaired electrons exist as shown in the dotted circles (Fig. 5.22a). It should be noted that the regions indicated by dotted circles in Fig. 5.22 correspond to those with unpaired electrons in the resonance structure (Fig. 5.19). Therefore, one could interpret that a small $|t_{ab}|$, which reduces $2^3E_{HL} - {}^1E_{HL}$ and thus contributes to meeting the energy level matching conditions [$2^3E_{HL} - {}^1E_{HL} \sim 0$ and $2^3E_{HL} - {}^1E_{HL} < 0$] for [$n$]acenes, is responsible for the diradical form appearing in the resonance structures because the unpaired electrons are localized well on the up and down zigzag edges, respectively, and thereby have small orbital overlaps due to the remote distance between them.

5.3.5 Summary

In this section, we have demonstrated the performance of the diradical character based design principle for singlet fission using the two-site diradical model as well as two kinds of typical alternant hydrocarbons, [n]acenes and [n]phenacenes. We found that the increase in the diradical character is induced by the increase in U ($=2K_{HL}$), which describes the concept of one of the conventional singlet fission design guidelines on alternant hydrocarbons, and/or of small $|t_{ab}|$. We also derived the analytical expression of one of the energy level matching conditions [$2E(T_1) - E(S_1) \sim 0$ or $2E(T_1) - E(S_1) < 0$] in the two-site diradical model. The dimensionless $[2E(T_1) - E(S_1)]/U$ ($=f$) in the energy level matching condition for singlet fission is clarified to be correlated with the diradical character (y_0) through the ratio between the S_1–T_1 gap (U) and the frontier orbital gap ($2|t_{ab}|$). Moreover, the best energy level matching ($f \sim 0$) is found to be satisfied for a quite weak but not negligible diradical character ($0.00 < y_0 \leq \sim 0.06$) within the two-site model. Importantly, this is a general result being independent from the molecular species. Therefore, we have concluded that a weak diradical character is an essential factor for a molecule exhibiting efficient singlet fission. The quantum chemical calculations on a series of [n]acenes and [n]phenacenes show that only some [n]acenes satisfy the energy level condition though [n]phenacenes do not, the difference of which originates in $|t_{ab}|$ rather than U in the present case, and is found to be distinguished by the diradical character. It is noted that the conventional design guideline choosing alternant hydrocarbons does not necessarily work for singlet fission as shown by [n]phenacenes. Moreover, on the basis of the comparison between the LNO distributions and the resonance structures, small $|t_{ab}|$ values of [n]acenes as compared to those of [n]phenacenes are found to correlate with the contribution of the diradical form in the resonance structures, the unpaired electrons of which should decrease $|t_{ab}|$ due to the latitudinal remote distance between the electrons occupied in the LNOs. These results demonstrate the reliability and advantage of the diradical character based design guideline for singlet fission molecules, and also clarify the importance of diradical forms in a resonance structure in terms of the small transfer integral.

References

1. A.S. Davydov, Soviet Phys. Uspekhi **82**, 145 (1964)
2. M. Kasha, H.R. Rawls, M. Ashraf El-Bayoumi, Pure Appl. Chem **11**, 371 (1965)
3. M.B. Smith, J. Michl, Chem. Rev. **110**, 6891 (2010)
4. S. Singh et al., J. Chem. Phys. **42**, 330 (1965)
5. M.C. Hanna, A.J. Nozik, J. Appl. Phys. **100**, 074510 (2006)
6. H. Najafov, B. Lee, Q. Zhou, L.C. Feldman, V. Podzorov, Nat. Mater. **9**, 938 (2010)
7. A. Rao, M.W.B. Wilson et al., J. Am. Chem. Soc. **132**, 12698 (2010)
8. P.J. Jadhav, P.R. Brown et al., Adv. Mater. **24**, 6169 (2012)
9. P.J. Jadhav et al., Nano Lett. **11**, 1495 (2011)
10. B. Ehrler, Nano Lett. **12**, 1053 (2012)
11. B. Ehrler et al., Appl. Phys. Lett. **101**, 113304 (2012)
12. S.T. Roberts, J. Am. Chem. Soc. **134**, 6388 (2012)
13. Y. Takeda, R. Katoh et al., J. Electron. Spectrosc. Relat. Phenom. **78**, 423 (1996)
14. E.C. Greyson, J. Vura-Weis et al., J. Phys. Chem. B **114**, 14168 (2010)
15. W.-L. Chan, M. Ligges, X.-Y. Zhu, Nat. Chem. **4**, 840 (2012)
16. C.A. Coulson, G.S. Rushbrooke, Proc. Cambridge. Phil. Soc. **36**, 193 (1940)
17. A. Akdag, Z. Havlas, J. Michl, J. Am. Chem. Soc. **134**, 14624 (2012)
18. S. Ito, T. Minami, M. Nakano, J. Phys. Chem. C **116**, 19729 (2012)
19. K. Schulten, M. Karplus, Chem. Phys. Lett. **14**, 305 (1972)
20. M. Nakano, H. Fukui, T. Minami et al., Theor. Chem. Acc. **130**, 711 (2011)
21. M. Nakano, H. Fukui, T. Minami et al., erratum **130**, 725 (2011)
22. M. Nakano et al., Phys. Rev. Lett. **99**, 033001 (2007)
23. K. Kamada et al., J. Phys. Chem. Lett. **1**, 937 (2010)
24. T. Minami, M. Nakano, J. Phys. Chem. Lett. **3**, 145 (2012)
25. M. Nakano, T. Minami et al., J. Chem. Phys. **136**, 0243151 (2012)
26. M.W. Schmidt et al., J. Comput. Chem. **14**, 1347 (1993)
27. H. Nagai, M. Nakano et al., Chem. Phys. Lett. **489**, 212 (2010)
28. Y. Kawashima, T. Hashimoto, H. Nakano et al., Theor. Chem. Acc. **102**, 49 (1999)
29. A.F. Schwerin, J.C. Johnson et al., J. Phys. Chem. A **114**, 1457 (2010)
30. H.A. Frank, J.A. Bautista et al., Biochemistry **39**, 2831 (2010)
31. A. Konishi, Y. Hirao et al., J. Am. Chem. Soc. **132**, 11021 (2010)
32. C. Lambert, Angew. Chem. Int. Ed. **50**, 1756 (2011)
33. K. Nakasuji, T. Kubo, Bull. Chem. Soc. Jpn **10**, 1791 (2004)
34. T. Kubo, A. Shimizu et al., Angew. Chem. Int. Ed. **44**, 6564 (2005)
35. T. Weil et al., Angew. Chem. Int. Ed. **49**, 9068 (2010)
36. E. Clar, Monatsh. Chem. **87**, 391 (1956)
37. A. Bohnen, Angew. Chem. Int. Ed. **29**, 525 (1990)
38. F.O. Holtrup, Chem. Eur. J. **3**, 219 (1997)
39. N.G. Pschirer, Angew. Chem. Int. Ed. **45**, 1401 (2006)
40. T. Minami et al., J. Phys. Chem. Lett. **3**, 2719 (2012)
41. R. Baer et al., Annu. Rev. Phys. Chem. **61**, 85 (2010)
42. T. Minami, M. Nakano, F. Castet et al., J Phys Chem Lett **2**, 1725 (2011)
43. I. Paci, J.C. Johnson et al., J. Am. Chem. Soc. **128**, 16546 (2006)
44. V.K. Thorsmølle, R.D. Averitt et al., Phys. Rev. Lett. **102**, 017401 (2009)
45. T. Minami, S. Ito, M. Nakano, J. Phys. Chem. Lett. **4**, 2133 (2013)
46. T. Minami, S. Ito, M. Nakano, J. Phys. Chem. A **117**, 2000 (2013)
47. S. Hirata, M. Head-Gordon, Chem. Phys. Lett. **314**, 291 (1999)
48. Y. Shao, M. Head-Gordon, A.I. Krylov, J. Chem. Phys. **118**, 4807 (2003)
49. A.D. Becke, Phys. Rev. A **38**, 3098 (1998)
50. C. Lee, W. Yang, P.G. Parr, Phys. Rev. B **37**, 785 (1988)
51. F. Wang, T. Ziegler, J. Chem. Phys. **121**, 12191 (2004)

52. Y. Tawada, T. Tsuneda et al., J. Phys. Chem. **120**, 8425 (2004)
53. S. Arulmozhiraja, M.L. Coote, J. Chem. Theor. Comput. **7**, 1296 (2011)
54. Y. Shao, L. Fusti-Molnar et al., *Q-CHEM Version 4.0* (Q-Chem. Inc, Pittsburgh, 2008)

Chapter 6
Summary and Future Prospects

Abstract The diradical character based design principles for efficient functional substances—highly efficient open-shell singlet nonlinear optical (NLO) systems and singlet fission (SF) molecules—are summarized. The remaining problems to be solved for designing open-shell singlet NLO and SF materials are presented together with future possible extension of the present concept.

Keywords Diradical character · Open-shell singlet · Molecular aggregate · Electron dynamics · Environment effect

In this book, we present novel guiding principles for controlling excitation energies and related properties—nonlinear optical (NLO) responses and singlet fission (SF)—in open-shell singlet molecular systems, which are really synthesized and some of them are found to be thermally stable. These novel principles are constructed based on the correlation between excitation energies/properties and diradical character y, which is a chemical index of "instability of chemical bond" (in other words, $1 - y$ implies the effective bond order) or a degree of electron correlation. One of the advantages of taking a diradical character viewpoint exists in the fact that the diradical character is a quantity concerning the effective chemical bond in the ground state, and thus is closely related to the conventional chemical concepts, e.g., covalency/ionicity, resonance structures, and aromaticity. Namely, once desired functionalities of molecules are described by the diradical character, we could employ the conventional chemical concepts or guidelines to tune the diradical character. Such tuning is much easier to perform than finding the direct control scheme of key physico-chemical parameters, e.g., excitation energy differences and transition properties between excited states.

As a guiding principle for highly efficient NLO molecular systems, we have found that systems with intermediate diradical character gives larger NLO properties, i.e., the first and second hyperpolarizabilities, than closed-shell and pure diradical systems. The intermediate diradical character turns out to be closely related to the aromaticity, edge shape, architecture and size of polycyclic aromatic hydrocarbons (PAHs), asymmtericity, donor/acceptor substitutions, transition metal–metal multiple bonds, etc. Although in general, molecules with intermediate

M. Nakano, *Excitation Energies and Properties of Open-Shell Singlet Molecules*,
SpringerBriefs in Electrical and Magnetic Properties of Atoms, Molecules, and Clusters,
DOI 10.1007/978-3-319-08120-5_6

diradical characters are thermally unstable, it is possible to realize highly efficient stable NLO molecules with intermediate diradical characters by kinetically stabilization due to the introduction of bulky substituents and by the introduction of donor/acceptor substitutions. In particular, the introduction of donor/acceptor substituents into the symmetric open-shell singlet systems causes asymmetric electronic distributions, which are found to provide remarkably larger enhancement of the NLO properties than symmetric open-shell singlet systems with intermediate and large diradical characters as shown in Chap. 3. There are several targets to tackle in open-shell singlet NLO systems:

(I-a) Supra/super-molecules with multi-radical characters
(I-b) Heavy main group compounds
(I-c) Effects of spin and charge states
(I-d) Intermolecular interaction effects in open-shell singlet aggregates
(I-e) Electron dynamics in open-shell singlet systems
(I-f) Vibrational effects on the open-shell singlet character

For (I-a), (I-c) and (I-d), we have to investigate aggregate models composed of open-shell monomers in order to clarify the effects of the variations in the multiradical character and charge/spin states on the ground- and excited-electronic states, and thus on the NLO responses. For (I-b), a weak hybridization of s and p orbitals tends to cause the open-shell singlet character in heavy main group compounds [1]. Such molecular systems are very intriguing from the viewpoint of novel NLO substances since they can realize a wide range of diradical characters in the ground state. For (I-e), a definition of diradical character could be extended to the dynamical behavior, which is useful for characterizing the electron dynamics and for controlling the effective chemical bonds, i.e., electron correlations, using superposition states between the ground and excited states. For (I-f), structural changes due to the vibrational modes may affect the diradical character and *vise versa*. Indeed, some changes in the Raman spectra caused by the change in the relative contribution of benzenoid (open-shell singlet) and quinoid (closed-shell) structures were observed in PAHs of different size [2–4].

For singlet fission (SF), the electronic structures of organic open-shell singlet molecules are discussed using the valence configuration interaction (VCI) model and several quantum chemical calculations in order to reveal the correlation between essential excitation energies concerning SF and singlet open-shell characters. On the basis of the results, we have newly proposed a novel molecular design principle for SF based on "diradical character view of singlet fission", which predicts that weak/intermediate diradical molecules exhibit efficient SF. According to this principle, several polycyclic aromatic hydrocarbons PAHs such as terrylene, bisanthene, and zethrene were predicted to be the promising candidates for SF. Although this guideline has provided one of the bases of molecular design for SF, there are still much room for investigation of detailed design principles of excited states (absolute value of excitation energy, character of wavefunction, and so on) for real molecules as well as of the clarification of SF dynamics in solid state. To further develop the design principles of molecules, the following issues to be solved are presented as the future prospects.

(II-a) SF molecule with high triplet excitation energy
(II-b) Molecular design for class II molecules
(II-c) Non-adiabatic SF dynamics due to the vibrational coupling with related excited electronic states

The importance of (II-a) is described as follows. Although the mechanism of exciton dissociation has not been clarified in organic photovoltaic cells (OPVs), it is believed that an efficient exciton dissociation is driven by some amount of excess energy, which is defined by the difference in free energy between thermalized exciton and dissociated charges [5, 6]. The amount of excess energy for maximizing the yield of free charge generation was indeed found to be about 0.5 eV for the polymer/fullerene blend film [7, 8]. For OPVs utilizing SF, the excess energy relates to the free energy of triplet exciton. This indicates that a high triplet excitation energy is required to cause an efficient exciton dissociation for OPV utilizing SF. However, it seems not to be easy to increase the triplet excitation energy of SF molecule because, a candidate of singlet fission is predicted as a weak/intermediate diradical molecule, the triplet excitation energy of which is generally small. We therefore need to construct a molecular design guideline to satisfy both the requirements of high triplet excitation energy as well as weak/intermediate diradical character. A possible strategy will be proposed by tuning diradcial character through detailed analysis of structural dependences of effective Coulomb repulsion U and transfer integral t [9]. For (II-b), since few molecules belong to class II, in which $S_1 =$ (HOMO → LUMO + 1) or (HOMO − 1 → LUMO) singly excited state [10] and little is known for the mechanism of SF in class II molecules, a comprehensive understanding of the SF mechanism in class II molecules has the possibility of extending the range of potential candidates for efficient SF. Furthermore, to evaluate the efficiency of SF, we have to investigate the SF dynamics in solid states (II-c), where the non-adiabatic coupling between one exciton state generated by photoabsorption and a dark state (composed of two coupled triplet states) plays an important role to achieve rapid decomposition into two triplet excitons. In addition to tuning the relative excitation energies of a molecule with the equilibrium geometry to satisfy the energy level matching conditions, the configuration of molecules in solid state and the coupling between electronic states with intra- and/or inter-molecular vibrations are predicted to affect the efficiency of the non-adiabatic coupling causing the SF. To this end, we have to develop the reliable calculation methods for related excited electronic states and vibrational modes for appropriate cluster models composed of monomers with singlet open-shell character in solid states, where the environment effects on the target cluster systems should be effectively taken into account. In a systematic computational design for SF systems, for candidate molecules screened by the diradical character based design principle for SF, the SF dynamics is investigated to clarify their performance in SF dynamics. Furthermore, such detailed analysis of SF mechanism will be useful for understanding the opposite process, i.e., triplet-triplet annihilation, which realizes photon upconversion (UC) [11, 12]. Unravelling the control scheme of these two processes, SF and UC, is a challenging theme in physical chemistry and materials science.

Although the concept of "open-shell singlet", which is characterized by "diradical character", was presented about 40 years ago, we have theoretically clarified the correlations of diradical character with molecular structures, reactivities, and ground and excitation energies/properties. On the basis of these correlations, we can obtain deeper understanding of effective chemical bonds as well as novel design principles for highly active functional, e.g., optical and magnetic, molecular systems. In particular, an intermediate diradical character region is essential for controlling/enhancing the functionalities by slight external chemical/physical perturbations through molecular structural change, donor/acceptor substitution, electric field application, modification of solvents or crystal field, and more generally, reservoir engineering (environment tuning). Recent rapid developments of synthetic and physical methods for preparing thermally stable molecular systems with a wide range of diradical characters lead to realizing various "open-shell singlet systems". Further experimental and theoretical studies on structures, reactivities, and functionalities of such novel open-shell molecular systems based on the "open-shell character" will be intensively desired from the broad viewpoints of science and engineering.

References

1. M.F. Breher, Coord. Chem. Rev. **251**, 1007 (2007)
2. Y. Li et al., J. Am. Chem. Soc. **134**, 14913 (2012)
3. Z. Zeng et al., J. Am. Chem. Soc. **135**, 6363 (2013)
4. Z. Sun et al., J. Am. Chem. Soc. **135**, 1829 (2013)
5. T.M. Clarke et al., Chem. Rev. **110**, 6736 (2010)
6. A.A. Bakulin et al., Science **335**, 1340 (2012)
7. H. Ohkita, S. Cook, Y. Astuti, W. Duffy, S. Tierney, W. Zhang, M. Heeny, I. McCulloch, J. Nelson, D.D.C. Bradley, J.R. Durrant, J. Am. Chem. Soc. **130**, 3030 (2008)
8. D.C. Coffey, B.W. Larson, A.W. Hains, J.B. Whitaker, N. Kopidakis, O.V. Boltalina, S.H. Strauss, G. Rumbles, J. Phys. Chem. C **116**, 8916 (2012)
9. T. Minami, S. Ito, M. Nakano, J. Phys. Chem. Lett. **4**, 2133 (2013)
10. M.B. Smith, J. Michl, Chem. Rev. **110**, 6891 (2010)
11. T.N. Singh-Rachford, F.N. Castellano, Coord. Chem. Rev. **254**, 2560 (2010)
12. J. Zhao, S. Ji, H. Guo, RSC Adv. **1**, 937 (2011)

Index

M. Nakano, *Excitation Energies and Properties of Open-Shell Singlet Molecules*,
SpringerBriefs in Electrical and Magnetic Properties of Atoms, Molecules, and Clusters,
DOI 10.1007/978-3-319-08120-5

CPSIA information can be obtained at www.ICGtesting.com
Printed in the USA
LVOW01s0129080814

398039LV00002B/11/P